"*The Shark's Paintbrush* reveals how nature is inspiring design to be more efficient, effective, resilient, and beautiful. In Nature's 3.8 billion years of design experience, the roughly 99% of designs that didn't work got recalled by the Manufacturer. The 1% that survived can teach profound lessons about how things should be made, how they work, and how they fit. Jay Harman's immersion in and curiosity about the natural world have made him one of the best biomimetic designers. *The Shark's Paintbrush* is a masterly field guide for all designers and entrepreneurs who aspire to help the world of the made work like, and live harmoniously with, the world of the born."
–Amory B. Lovins, Chairman and Chief Scientist, Rocky Mountain Institute

"At this point in our global ecological crisis, the survival of humanity will require a fundamental shift in our attitude toward nature: from finding out how we can dominate and manipulate nature to how we can learn from her. In this brilliant and hopeful book, Jay Harman shows us how far the new field of Biomimicry has already progressed toward this goal. *The Shark's Paintbrush* makes for fascinating and joyful reading —much needed in these dark times."
–Fritjof Capra, author of *The Science of Leonardo* and *The Tao of Physics*

"It blows my mind how the same patterns repeat themselves in flowers, growth in animals, and in shell spirals. When engineers mimic designs from nature, more efficient designs for windmills and fuel efficient vehicles are the result. Many of the breakthroughs are counter intuitive. Jay Harman has told the promise of Biomimicry in a compelling way. *The Shark's Paintbrush* is essential reading for people interested in design and engineering."
–Temple Grandin, author of *Thinking in Pictures*

"A useful update on recent developments in biomimicry and an intriguing case for innovative green technology that goes beyond sustainability."
–Kirkus Review

"The world's most elegant engineer is without a doubt Mother Nature. She's efficient, creative, and has time on her hands to put her innovations to the test, and Harman thinks we can learn a lot from watching her work. His company, PAX Scientific Inc., focuses on sustainable design solutions based on mimicking biological adaptations. Sounds far out, but the practice has been around for a while. Archimedes' screw, a type of water pump that is still used today, is an early example of taking a concept from nature—in this case the spiral—and putting it to practical use. The eponymous paintbrush is equally fascinating. German scientists developed a special paint that, when applied to the hull of a ship in such a way as to mimic the water-repellent design of sharkskin, reduces drag by 5%. This can result in savings of 2000 tons of fuel per ship, per year. Looking elsewhere, scientists are learning about anticoagulants from leeches, acoustics from dolphins, antibiotics from Komodo dragons, shock absorbers from woodpeckers, and computer networks from slime molds. Harman points out that many of these developments would not only save money, but also prove an enormous boon to the survival of Mother Nature. His vision of a biomimetic "new global economy" is timely, crucial, and thrillingly eye-opening."
–Publisher's Weekly

THE SHARK'S
PAINTBRUSH

THE SHARK'S PAINTBRUSH

BIOMIMICRY
AND HOW
NATURE
IS INSPIRING
INNOVATION

JAY HARMAN

White Cloud Press
Ashland, Oregon

White Cloud Press titles may be purchased for educational, business, or
sales promotional use. For information, please write:

Special Market Department
White Cloud Press
PO Box 3400, Ashland, OR 97520
Website: www.whitecloudpress.com

Jacket design by C Book Services
Jacket illustration © Gary Bell/
OceanwideImages.com

Library of Congress Cataloging-in-Publication
Harman, Jay.
The shark's paintbrush : biomimicry and how
nature is inspiring innovation / Jay Harman.
p.   cm.
1. Biomimicry.   I. Title.
T173.8.H373 2013
600— dc23
2012015185
ISBN 978-1-935952-84-8

MANUFACTURED IN THE UNITED STATES

1 3 5 7 9 10 8 6 4 2

First Edition

To the shareholders and staff of PAX, who took the risk of supporting a new paradigm and whose generosity, trust, patience, and hard work brought it to life; and especially to Marion Weber—a true angel.

Contents

. . . .

Introduction: The Nature of Innovation *1*

PART ONE: A NEW GOLDEN AGE

1: *The Next Industrial Revolution*..........................13

2: *Going with the Flow* 39

3: *Catching the Worldwide Wave*.................................65

PART TWO: BIOMIMICRY AT WORK

4: *Secrets from the Sea* 85

5: *Scales and Feathers* 113

6: *The Bee's Knees*................................. 137

7: *Spores and Seeds* 164

8: *Wampum*................................. 191

PART THREE: THE NATURE OF CHANGE

9: *The Corporate Jungle* 215

10: *Dollar Signs*..................................249

11: *Reorganizing (Your) Business*..........................264

Epilogue 290

Acknowledgments................................ 295

Notes 297

Index 329

THE SHARK'S
PAINTBRUSH

THE NATURE OF INNOVATION

Why does the bumblebee have better aerodynamics than a 747?

How can a seashell keep a microchip from overheating?

How can the colors of a butterfly's wing reduce the world's lighting energy bill by 80 percent?

How can fleas' knees and bees' shoulders help scientists formulate a near-perfect rubber?

How will the answers to these and similar questions forever change our lives?

Most young ladies sunning by the pool or beach probably aren't thinking about a hippopotamus, let alone its perspiration. However, it turns out that hippo sweat provides a highly effective, four-in-one sunblock. We humans perspire by allowing salt water to leave our pores, using the physics of evaporation to cool the skin. Hippos—long-lost cousins of whales and dolphins—solve more than just a cooling problem by secreting a blend of chemicals that takes care of many challenges simultaneously. Besides being an excellent, nontoxic sunscreen (though perhaps a little aromatic in its natural form), hippo sweat is also antiseptic, insect repelling, and antifungal.

Researchers at Kyoto Pharmaceutical University and the University of Trieste, as well as Dr. Christopher Viney and his materials engineering team at the University of California, Merced, have studied the rust-colored combination of mucus and chemicals secreted by hippos. They found two pigments that absorb light across the ultraviolet-visible range, with crystalline structures that ensure the material spreads over the entire skin without the need for being rubbed on by hand (a challenge for a hippo). The pigments turn white skin a shade darker, while simultaneously slowing the rate of bacteria growth.

With one million Americans developing skin cancers each year, the market for sunscreen is substantial—$640 million a year and growing. Yet of the more than eighteen hundred products containing sunscreen on the market today, three out of four have been shown not to live up to their claims. In addition, most chemicals applied to the skin can be absorbed straight into the bloodstream, so some scientists are concerned that certain sunscreens may prevent sunburn but introduce toxins that can still cause cancer. In fact, the FDA recently responded to concerns about product efficacy by tightening the rules for sunscreen labels.

Of course, few bathing beauties would find delight in smelling like a hippo's armpit, so there's the exciting opportunity to synthesize the sweat's beneficial properties, having it smell more like, say, coconuts. Longer term, scientists anticipate applying the hippo's chemistry to exterior paints, clothing, and other UV-sensitive products.

Whether finding inspiration on hippos to reduce skin cancer or developing better road systems by studying the tracks made by slime molds seeking food, biomimicry, or bio-inspiration as some call it, very simply means applying lessons learned from nature to solve human problems. Examples of biomimicry include everything from energy-producing solar cells that mimic tree leaves to lifesaving pharmaceutical breakthroughs based on the biology of lizards to antibacterial paints that emulate sharkskin to highly profitable businesses that improve their organizational structures based on redwood groves. Even the Velcro you undoubtedly have somewhere in your closet is a prime example of biomimicry in action.

Why do we need biomimicry right now? Despite the confusing claims and counterclaims of scientists, corporations, interest groups, and politicians about whether the earth is in catastrophic decline, we all know in our hearts that something is not right. Half of all humans live on less than $2.50 per day. Fuel prices are unstable. Weather is rapidly growing more severe and unpredictable. We're living through a mass extinction of species. In the wake of the worldwide financial meltdown, innovation has slowed. In medicine, we're losing battles with antibiotic-resistant strains of bacteria, while the incidence of cancer, Alzheimer's, autism, and diabetes is escalating. And, despite living in a period with more trained researchers, engineers, and doctors than in all of history combined, the future viability of our race is in increasing danger.

The story doesn't need to end this way. Most of our environmental and economic problems result from an out-of-date way of doing business. Industry has continued to depend on the same old "heat, beat, and treat" methods that were mechanized in the industrial revolution, but these methods simply aren't sustainable. Nature, on the other hand, constantly evolves, survives, and thrives, while not using up or endangering its base resources. It reinvents itself, adapting and beginning anew with irrepressible optimism.

As a serial entrepreneur and inventor, I've spent the past thirty years starting and growing multimillion-dollar research and manufacturing companies that develop, patent, and license innovative products, ranging from prize-winning watercraft to interlocking building bricks to electronic information systems and noninvasive technology for measuring blood glucose and other electrolytes. Now I find myself credited with being among the first scientists to make biomimicry a cornerstone of modern and future engineering. My latest ventures—PAX Scientific and its subsidiary companies—design more-energy-efficient industrial equipment, including refrigeration, turbines, fans, mixer systems, and pumps based on nature's fluid flow geometries.

Simply stated, I'm on a mission to halve the world's energy use and greenhouse gas emissions through biomimicry and the elimination of waste. I'm also on a mission to inspire others to climb on board a new wave of possibility and optimism that is rapidly gaining momentum: All over the world, across dozens of industries, people are finding profitable solutions to seemingly intractable problems by partnering with nature. This book will clearly demonstrate that nature is the best source of answers to the technological, biological, and design challenges that we face as humans.

Scientists have already identified more than two million species of life on earth; some estimate that there may be as many as one hundred million. Each one has evolved hundreds of optimized solutions to life's challenges, many of which can be readily applied to the very problems facing human enterprise and survival. By constantly creating conditions conducive to life, with zero waste and a balanced use of resources, nature is clean, green, and sustainable. Following nature's design mastery, we *can* achieve greater wealth and economic sustainability. We can do this without sacrifice, while protecting our planet. How biomimicry is invigorating current business models, and how individuals and companies can reap the rewards that this burgeoning industry has to offer, is exactly what this book is all about.

Looking back, I had long shown the makings of a biomimic. When I was a boy growing up in Australia, I knew that fish were highly effective swimmers. They usually survived my admittedly ungainly attempts to catch them with a spear that I'd made from a broomstick and bent nails. Hoping to paddle farther out to good fishing sites without getting tired, I experimented with hammering the sides and bottom of my homemade metal canoe into the shapes I'd seen on fish and ducks. Of course, my canoe became easier to paddle, though I'm not sure if that could have been empirically measured or just seemed that way to my biased enthusiasm.

Regardless of efficacy, I was captivated and convinced that I was on the right path. These observations and experiments launched me into a lifelong career, first as a naturalist, observing nature's exquisitely evolved shapes, and later as an inventor, adapting those shapes to design more efficient industrial devices. During the first half of my business career, I built and sold award-winning products and companies in Australia, the United Kingdom, and the United States, but it wasn't until the late 1990s that I realized I was part of an emerging scientific discipline.

From the Greek *bios*, meaning "life," and *mimesis*, "to imitate," the term *biomimicry* was first coined in 1997 by Janine Benyus, the gifted naturalist, educator, and author of the landmark book *Biomimicry*. But biomimicry isn't new. Humans have copied nature for millennia, with varying degrees of accuracy and understanding. Our early human ancestors borrowed solutions from the animals and plants they saw around them. Seals swimming below arctic ice create and maintain holes through which they can surface to breathe; Inuit hunters mimicked the way polar bears lie in wait beside those breathing holes to catch a rich, blubber dinner. Polynesian outrigger canoes' design echoed that of floating seed pods. Aboriginal Australians even mimicked bird wings with their boomerangs. Certain shapes and tools were repeated around the world, created by people who were separated by vast geographical distances yet simultaneously immersed in and observing nature's problem-solving strategies.

Within just a few thousand years—a millisecond in evolutionary time—humans had developed much more complex tools, and the intellectual theories to support them. Newtonian physics, the industrial revolution, and the nineteenth century age of enlightenment spurred tremendous technological development and transformed our social mores. A consequence of this paradigm shift, however, was that humanity's view of the world changed from an organic to a mechanistic one. Early engineers saw the potential of breaking up any system into components and rearranging the parts. Innovations in machinery and materials led to mass production: making thousands and then millions of exactly the same forms out of flat metal plates and square building blocks. However, for all its positive impact on the economics and culture of the era, the industrial revolution's orientation was shortsighted. In the rush to understand the world as a clockwork mechanism of discrete components, nature's design genius was left behind—and with it the blueprints for natural, nontoxic, streamlined efficiency. A new set of values emerged, such that anything drawn from nature was dismissed as primitive in favor of human invention. Just as the pharmacology of the rain forests, known to indigenous people for millennia, has been largely lost to modern science, so too were the simple rules of natural design obfuscated. As our societies became more urban, we went from living and working in nature and being intimately connected with its systems, to viewing nature as a mere warehouse (some might say, whorehouse) of raw materials waiting to be plundered for industrial development.

The industrial revolution was also about cheap and plentiful power. If you needed more speed, you didn't look to nature to find a more efficient way, you just shoveled in more fuel and blasted your way forward. That approach worked well enough until the side effects began mounting: polluted air and water, stripped lands, diminishing access to cheap fossil fuel, new public health risks, and global warming.

Nature works on an entirely different principle. Its mandate for survival is to use the least amount of material and energy to get the job done—the job being to survive and re-create itself without damaging its foundational ecosystem. It doesn't stamp out flat plates; it doesn't create straight lines. For example, the ultraefficient human cardiovascular system has sixty thousand miles of plumbing, yet there's not a straight pipe inside. However, it is beyond compare when it comes to energy efficiency. How many machines can drive anything sixty thousand miles on one-and-a-half watts of power? That's less than the power consumed by many bedroom night-lights.

Given the mounting side effects of our wasteful use of energy, the imperative and opportunity to create a new global economy is upon us. We must leap into a new business and technology model or go the way of the dinosaurs. The opportunity starts with embracing nature's phenomenal efficiency and functionality. From nature's point of view, there is no energy shortage—never has been and never will be. Our whole universe and everything in it is made of energy. In nature, survival of a species depends on its optimal use of energy. If we study and faithfully copy nature's strategies for energy use, we can avert the developed world's escalating energy crisis—a crisis that is already entrenched for two-thirds of the earth's people. After life's 3.8 billion years of trial and error, experimentation, and a limitless research budget, the time has come for us to turn to nature's vast library of elegant, efficient methodologies, freely available to those who ask the right questions.

In the first part of this book—A New Golden Age—I'll introduce the tremendous potential for modern biomimicry. In one study of worldwide patent databases, bio-inspired inventions grew by a factor of ninety-three between 1985 and 2005. The rate of growth has only increased since then. Some believe that the benefits offered by biomimicry are so great, compared to conventional technologies, that bio-inspired design will replace old methods completely within the next thirty years. That's a lot of opportunity, and a lot of honest money to be made for the sake of a healthier planet—and yet, few people and businesses are even aware of the term biomimicry.

The basis of biomimicry's contemporary applicability is simple and profound: If a plant or animal had an effective solution, it survived and over time became ever more adapted to its niche. Now, with the increasing sophistication of our scientific devices, we can more precisely study nature's strategies and adapt them to solve our most intractable problems.

One of the best-known examples of commercially successful biomimicry is Velcro. When hiking in the Alps in 1941, a Swiss inventor named George de Mestral became annoyed by the repeating problem of burrs sticking to his socks and his dog's fur. He looked at the annoyances under a microscope and discovered the hook-and-loop structure that became the basis for Velcro.

Modern biomimicry is far more than just copying nature's shapes. It includes systematic design and problem-solving processes, which are now being refined by scientists and engineers in universities and institutes worldwide.

The first step in any of these processes is to clearly define the challenge we're trying to solve. Then we can determine whether the problem is related to form, function, or ecosystem. Next, we ask what plant, animal, or natural process solves a similar problem most effectively. For example, engineers trying to design a camera lens with the widest viewing angle possible found inspiration in the eyes of bees, which can see an incredible five-sixths of the way, or three hundred degrees, around their heads.

The process can also work in reverse, where the exceptional strategies of a plant, animal, or ecosystem are recognized and reverse engineered. De Mestral's study of the tenacious grip of burrs on his socks is an early example of reverse engineering a natural winner, while researchers' fascination at the way geckos can hang upside down from the ceiling or climb vertical windows has now resulted in innovative adhesives and bandages.

Designs based on biomimicry offer a range of economic benefits. Because nature has carried out trillions of parallel, competitive experiments for millions of years, its successful designs are dramatically more energy efficient than the inventions we've created in the past couple of hundred years. Nature builds only with locally derived materials, so it uses little transport energy. Its designs can be less expensive to manufacture than traditional approaches, because nature doesn't waste materials. For example, the exciting new engineering frontier of nanotechnology mirrors nature's manufacturing principles by building devices one molecule at a time. This means no offcuts or excess. Nature can't afford to poison itself, either, so it creates and combines chemicals in a way that is nontoxic to its ecosystems. Green chemistry is a branch of biomimicry that uses this do-no-harm principle, to develop everything from medicines to cleaning products to industrial molecules that are safe by design. Learning from the way nature handles materials also allows one of our companies, PaxFan, to build fans that are smaller and lighter while giving higher performance. Finally, nature has methods to recycle absolutely everything that it creates. In nature's closed loop of survival on this planet, everything is a resource and everything is recycled—one of the most fundamental components of sustainability. For all these reasons, as I heard one prominent venture capitalist declare, biomimicry will be *the* business of the twenty-first century. The global force of this emerging and fascinating field is undeniable and building on all societal levels.

In the second part of this book—Biomimicry at Work—you'll meet some of the remarkable animals and plants, along with the devoted scientists and engineers, who are proving the potential for biomimicry right now. From sharks, whales, and dolphins to lizards and leeches to bees and butterflies to trees and seashells, there are thousands of species already teaching us about engineering, chemistry, materials science, fluid dynamics, nanotechnology, medical devices, and on and on. Moreover, even the remains of extinct species can inform us. Think about it—millions of living species and ten billion life-forms in the past, each with unique solutions for us to capitalize on. As a naturalist, I've had a few adventures in the wild over the years, so I'll share some of my own encounters with biomimetic movers and shakers to introduce their secrets.

The last section of the book—The Nature of Change—explores the three chief principles of running a bio-inspired business and shares some of the challenges to their successful implementation. Throughout my career, I've seen some unnatural behavior in the boardroom as well as in the engineering lab. Rather than a walk in a park, the path to commercial success has been more of a trek through a corporate jungle. Now the path is clearing and bio-inspired design is coming out of the woods. I've watched as large companies, cities, and governments take steps to invest in biomimetic innovation and learn to tread more lightly on our planet. I'll provide pointers on how to run a business—whether large or small—more biomimetically and introduce you to biomimicry career opportunities in business and science. I'll also share practical instructions on how to develop and launch bio-inspired products, including personal stories of obstacles my companies have faced and how we overcame them.

This is a very exciting time in science and technology. I am inspired daily by the potential of applying nature's lessons to design a new golden age for the earth and for humanity—a golden age that is not only possible but realistically achievable. Biomimicry will get us there. Whether you're a CEO, corporate employee, commercial manufacturer, entrepreneur, politician, small business owner, college student looking to start a company, teacher wanting to share a world of positive choices with your students, or merely someone curious about this new paradigm, my hope is that one message rings loud and clear: By learning from nature, we can create more abundant, healthy, satisfying lives for ourselves, our children, and our planet.

Part One

....

A NEW GOLDEN AGE

Chapter 1

THE NEXT INDUSTRIAL REVOLUTION

My chest was bursting. I was thirty-five feet deep, swimming along a ten-foot-wide trench of white sand with limestone and coral sides. Rapidly increasing ocean surge rushed intermittently up the canyon, trying to flush me onto the shallowing sharp corals behind me. I had run out of breath and needed to get back to the surface.

I was diving in one of the most treacherous, and certainly most inaccessible, sites of my life—under the shadows of overhanging, sixty-feet high, sheer walls of jagged, fossilized coral. While the water here was only about forty feet deep, just one hundred yards from me the ocean floor dropped precipitously to a depth of seventeen hundred feet. I guessed I was the first person ever to swim voluntarily in this hellacious location. Although the cliffs were pounded by large waves for most of the year, this was one of a handful of days when the ocean was relatively calm and the water clear enough for me to embark on a long-anticipated, skin-diving expedition. I was a fit forty-two-year-old, but this dive was demanding all my power.

Wave surges were now kicking sand up from the bottom—reducing visibility. I had been surrounded by clouds of bright-colored tropical fish, often so thick that I couldn't see beyond them. A little disconcerting really, since sea mounts like this were preferred habitats for several species of man-eating sharks. I couldn't see them, but I know that any shark within a mile of me knew I was there. With the swell building so rapidly, I was suspended in a soup of appetizers with the possibility of lurking predators attacking from the abyss as well as at increasing risk of being thrown onto and scraped over sharp and jagged rock.

As I turned for the surface, my eye was caught by a fragment of white and blue pottery lying on the sand, its edges worn smooth. A sharp thrill moved through me. Dutch trading ships, which sailed near these waters in the seventeenth century, often carried Chinese porcelain—and there seemed no other reason for this piece to be in such a remote place. I grabbed the shard and tucked it into the sleeve of my wet suit. Fighting the growing push of the waves, I scanned the sea bed. After years of snorkeling and diving, I knew that there are no straight-sided shapes naturally occurring in the ocean, so my heart jumped when I saw a pale orange and brown, rectangular shape protruding from the reef just above the sand. It was the size and shape of a small, flat, house brick. I suppressed the pressing urge to get to the surface for air and pulled at the object.

Along with heavy chests of silver bullion, Dutch East Indies ships used small clay bricks, made in the Netherlands, as ballast to stabilize their ships on the long voyage to the Spice Islands, now known as Indonesia. Once there, the bricks were sold for use in buildings, and the silver— highly prized by local master silversmiths—was traded for spices. Worth their weight in gold, these condiments tantalized the taste buds of Europeans jaded by bland diets of salted meat and pickled or dried fish— not to mention disguising the taste of spoiled food in the days before refrigeration.

My brick looked similar to ballast I had seen on other wrecks. I pulled hard—it didn't budge. It had obviously been there for a very long time; coral had grown around it and cemented it to the reef. A strong surge suddenly shoved me hard up the channel toward the cliffs. I turned and swam with all my strength toward deeper, safer, water while twisting upward to fin to the surface. Once there, I blew hard to expel the water from my snorkel and gasped in fresh air. I rested for a couple of minutes, breathing deeply, replenishing my blood oxygen. My heart was thumping from the exertion and excitement. Could I have really found a treasure ship?

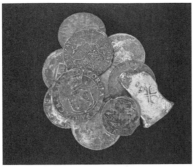

Silver treasure recovered from Dutch East Indies shipwrecks

Seventeenth-century ballast brick found on Indian Ocean dive

I had spent years looking for two missing Dutch East Indian ship-wrecks from the 1600s—the *Fortuin* and the *Aagtekerke*. The Dutch traders of that time left impeccable records of their voyages from Europe to Batavia (modern-day Jakarta), capital of the Spice Islands. As they built a trading empire, the Dutch amassed huge fortunes and created the largest private navy in history. Their ships sailed down the west coast of Africa, around the Cape of Good Hope, and across the lower reaches of the Indian Ocean. Here trade winds, known as the roaring forties, hurtled them forward at four miles per hour for thousands of miles to the desolate west coast of Australia, a vast and wild land still not explored or settled by Europeans. There they turned north for the last two thousand miles to Batavia.

Unfortunately, several silver-laden ships didn't turn left in time and banged into Australia, wrecking on its rugged coastline. At that stage in history, sailors could calculate only latitude, so they knew where they were relative to north and south, but how far east or west they had sailed was left to guesstimates. It wasn't until the invention of the ship's chronometer in 1735 that longitude could be determined.

I had dived on several Dutch shipwrecks: the *Batavia*, the *Zuytdorp*, the *Zeewijk*, and the *Vergulde Draeck* ("Gilt Dragon"). Each had carried large chests of silver coins weighing many tons. The Dutch at that time were able to salvage some coins from a couple of the wrecks, while the bulk of the remainder has now been retrieved by the Western Australian Museum. However, the *Fortuin* and the *Aagtekerke* disappeared without a trace. As these were fine, seaworthy ships with competent crews, it's unlikely that they would have simply foundered at sea. It was much more likely that they had crashed into a reef or cliff in a storm on a dark night. A combination of research, deduction, and intuition had brought me to this tiny corner where ocean met land. I was excited—convinced that this was the resting place of at least one of the missing ships. The shape and location of the sand channel I was swimming in was also suggestive of a wreck site. The Dutch flagship *Batavia* had carved out such a trench when it sank on an Australian coastal reef in 1629.

The waves were building—the window of calm weather was over. I took a deep breath and headed for the bottom again. The channel was now filled with clouds of sand at each wave surge; and the water, crashing onto the cliffs, was turning white with air bubbles. With little visibility and the waves washing back and forth, I had trouble getting to the site I had just left, but then I spotted the brick again. I grabbed hold of it, pulled my knife from the sheath strapped to my leg, and pounded at the encasing rock. Pieces chipped away, and then a particularly large surge yanked me toward the cliff. I hung on to the brick with everything I had. With the extra force of the pulling wave on my body, I wrenched my prize out of the rock, revealing blackened sand and coral underneath.

From previous wreck dives I knew that the black color, uncommon in coral environments, could be residue from oxidized silver. Could there be coins? It's hard to express my level of excitement as I tucked the brick into the jacket of my wet suit and swam back to the surface for air. I had just gotten to the bottom again when another surging wave thrust and dragged me along the channel's rocky side, ripping the hip of my wet suit and the back of my hand. This was getting much too dangerous.I had to get out of the water and climb the cliff back to safety.

I floated for several minutes, watching how each swell surged onto the fallen slabs of rock under the cliff. When I thought I understood the ocean's rhythm, I rode one of the waves onto the rock and landed several feet above the sea's low point. As fast as I could, I pulled off my fins and mask and clambered higher up the rocks. Another, larger wave swept me off my feet and tore my fins out of my hands. I grabbed at higher rocks and pulled myself up above the surf line—safe at last but very disappointed. After my swim, the weather closed in, and I've never had the opportunity to dive there again. A few years later, a vessel foundered a few miles away in rough weather and rescuers were unable to get into the area to save the passengers. Sadly, more than thirty people drowned, confirming the danger of this treacherous place.

THE BUSINESS OF BIOMIMICRY: A BLUEPRINT FOR A CLEAN, GREEN, AND SUSTAINABLE WORLD

Treasure hunting and the prospect of finding fabulous riches enthralls and excites us—though only a handful of people throughout history actually make such finds. What if there were incredible hidden treasures and wealth-creating opportunities that anyone could find and everyone could benefit from—just sitting and waiting for us to discover? There are, and the secrets are being unlocked as we speak. The most exciting part is that anyone—regardless of education, culture, or life circumstance—can make a new biomimicry discovery or imagine a valuable new application for life's hidden design mastery

In the past hundred years, we've created technological wonders that would have left our great-grandparents dumbfounded. Imagine Edison using a smartphone or the Wright brothers sitting in a NASA control room watching a spacecraft land on Mars. We have achieved comfort and security for millions, with more people today having health care, cars, home heating, electricity, televisions, sheets, and dishes, than the world's entire population in the year 1900. Unfortunately, to achieve this standard of living for one quarter of humanity, we are decimating the natural environment and have left more than half the earth's people in poverty. Future trends look dire. Media sources are replete with news of one environmental or social disaster after another—melting glaciers, mass extinctions, devastating droughts and floods, diminishing water reserves, the looming energy crisis, precipitous declines in oceanic fish stocks, desertification of farmlands, record wildfires, and vast oil spills. Are we rolling unstoppably toward certain catastrophe? Or are we surrounded by more opportunities than in all of history combined?

In the face of these worldwide challenges, there is a new frontier that offers tremendous opportunity without sacrifice. It's where I've been working and building multimillion-dollar businesses. It's a new paradigm—a new gold rush: biomimicry. Introduced by pioneering author and naturalist Janine Benyus in 1997, the term *biomimicry* refers to the rapidly growing discipline that identifies, analyzes, and adapts natural strategies to solve technological problems. Biomimicry is not just design that imitates or copies nature. It's design that asks the right questions in order to understand the mechanisms in nature's cornucopia of solutions, then uses that understanding to remedy problems—without creating new ones. You can start with an observation in nature and apply it, or start with a technological need and find a champion adapter in nature.

In the natural world, everybody eats somebody. For the 3.8 billion years since life emerged on earth, there has been a huge imperative for predator and prey to evolve survival strategies to the highest degree possible. Life depends on it. Now humanity's survival depends on it. As a naturalist, diver, and boat captain, I've studied and experienced many natural wonders close-up, from thousands of acres of bioluminescence in the ocean, so bright that it can be seen from space, to the superfast peregrine falcon, which has been clocked at almost 200 miles per hour, to the soaring leap of a spinner dolphin, which engineering calculations still can't fully explain. I learned early on that nature routinely exhibits dazzling feats of engineering that are far superior to human achievement.

Wild nature is full of stunning beauty. It is also ruthless and not often kind nor sweet nor gentle. Nature is no stranger to dealing with population explosions. We see it over and over throughout the natural world—plagues of locusts, mice, or starfish—huge population growth followed by nature's heavy-handed equalizer—starvation and disease. All species experience this and humans have been no exception. Throughout recorded history, we know of population meltdowns in Ethiopia, Russia, and Ireland, to name just a few. Archaeologists and anthropologists have also identified near extinctions of humans before cultural records began. The widely endorsed Toba catastrophe theory, for example, postulates that there was a cataclysmic eruption of a volcano seventy thousand years ago in what is now Indonesia. It caused a worldwide volcanic winter with a loss of summer growing seasons for six to ten years. Subsequent starvation plummeted the total human population to as few as one thousand breeding pairs. But life is tenacious, compelled to overcome obstacles and threats to its existence. Each individual is equipped with powerful survival instincts; whole species have even stronger strategies—because nature seems less concerned with individuals than the whole genus. Bugs mutate to survive pesticides. Termites chew through lead pipe to get at the tasty plastic insulation on underground electric cables. Bacteria become antibiotic resistant. No matter what we do, nature is a survivor—if necessary, at our expense.

Somehow humanity has determined itself to be special, impervious, with an inalienable right to be alive, yet across the entire natural world it is impossible to find evidence of a right to life in any species. Whatever right there is must be earned by adaptability and irrepressibility. The human species, now seven billion strong, is facing the greatest challenges to its survival in seventy thousand years. We humans, ourselves part and product of nature, can turn to the millions of successful adapters surrounding us to not only adapt and survive but also to thrive.

Nature designs, births, breathes, and sustains us. When we ignore or destroy nature, we destroy our very foundation. With our human intellect and bio-inspiration, however, we can choose to stand firmly on nature's foundation and flower as a species.

HOW BIG IS THE POTENTIAL?

Although the modern discipline of biomimicry is only about fifteen years old, bio-inspired products have already generated billions of dollars in sales. Among these are carpet tiles modeled on a forest floor; self-cleaning buildings inspired by the leaves of the lotus plant; and fabrics, paints, and cosmetics that derive their brilliance from the way color is created on peacock feathers. By using a material that mimics the way mussel shells maintain their grip on rocks, a new kind of plywood is being manufactured that eliminates toxic adhesives. New biomimetic products are coming online with big backing: Qualcomm, a world leader in wireless technology, invested seven years in development and nearly $1 billion on a factory to mass-produce their novel electronics screen inspired by the crystalline structure in butterfly wings. Bio-inspired products often see annual doubling in sales when they enter the market. They offer customers better performance, reduced energy requirements, less waste, and less toxicity, while being sold at prices competitive with existing products.

More than $200 billion was invested in sustainable businesses worldwide in 2010—a 40 percent increase from 2009—even as many other sectors sagged in the worst recession in eighty years. Biomimicry is sometimes described as a discipline of sustainable engineering, but any truly sustainable product or business is inherently biomimetic. Biomimicry creates products based on nature's peak achievers—all of whom are sustainable.

The growth potential is clear. Projections are that by 2025, biomimicry could represent $1 trillion of gross domestic product, including $300 billion of U.S. GDP. I know of bio-inspired technologies in pharmacology, water treatment, heat management, and refrigeration that have this potential alone. Another $50 billion could be counted, just in the United States, from the consequent reduction of carbon dioxide pollution and preservation of natural resources. It's estimated that 1.6 million U.S. jobs could be created by biomimetic businesses in the next fifteen years.

Since the beginning of the industrial revolution, business and environmental interests have traditionally been on opposite sides of the bottom line. While the occasional CEO might support initiatives that matched his or her own personal values, companies have generally adopted environmental principles only to comply with regulations, earn subsidies, or make marketing points with their customers. Tax breaks have been used to introduce cleaner technology into the marketplace, but these incentives tend to dry up when economic times get tough. In 1994, a British organization called SustainAbility introduced a model called the triple bottom line. The model is often subtitled "people, planet, and profits." The goal was to focus attention not just on economic return on investment but also on environmental and social values. While triple bottom line has been a useful business concept, it hasn't fundamentally changed the conditions that support or restrain ecofriendly innovation.

The Fermanian Business & Economic Institute (FBEI) at Point Loma Nazarene University has set out to change that outcome. The San Diego Zoo sponsored FBEI to produce the first economic impact report on the field, which was published in late 2010. The report concludes that by 2025, biomimicry could conservatively impact 15 percent of all chemical manufacturing, waste management, and remediation services. In the same period, 10 percent of all architecture and engineering services, textile production, transportation equipment, and utilities may be informed by biomimicry, along with 5 percent of food products, construction, plastics production, and computer equipment and information services.

Following the strong response to its economic impact report, FBEI released a new business model, called E2. Unlike the triple bottom line model, which adds or subtracts value based on a business's positive or destructive impact on the environment or social sector, the E2 model views economic and environmental interests as mutually supportive. In August 2011, FBEI also announced the Da Vinci Index, the first formal measure of biomimicry activity. Like the consumer price index, the Da Vinci Index combines data to show trends in biomimetic activity. Currently based primarily on U.S. statistics, FBEI started from the year 2000 to count biomimicry-related scholarly articles, patents, grants—and the dollar value of those grants.

As Lynn Reaser, chief economist of FBEI, said when the index was announced, "Two purposes of the index are to spread awareness of the field to investors, companies, policy makers, and universities and to begin monitoring that spread through real numbers. I personally believe that we are seeing progress in a truly game-changing technology."

The Da Vinci Index shows that:

- Biomimicry activity expanded more than seven and a half times between 2000 and 2010, with a compound growth rate of 22 percent.

- In 2009, more than 900 patent applications submitted to the U.S. Patent Office contained the words *biomimicry, bio-inspired,* or similar terms. Because it takes a number of years to complete the patenting process, the number of patent applications each year outstrips the number of patents granted. The number of U.S. patents granted that included biomimetic terms went from three in the year 2000 to forty-one in 2010. Moreover, a patent may be bio-inspired but not include any specific references to the biomimicry terms that were searched for by the Da Vinci Index, so the researchers are confident that these are conservative numbers.

- Scholarly articles on biomimicry, which are strong precursors of technology transfer to industry, multiplied five times in ten years. In 2010, there were more than fifteen hundred scientific papers related to biomimicry published worldwide, with nearly a third published in the United States. European and Asian universities are increasingly active, with entire research centers that are dedicated to either bio-inspired design or green chemistry, which undertakes to create nontoxic molecules that are as or more effective than competitor products for the same cost. Chemistry, material sciences, and engineering represent the majority of scholarly articles, followed by physics, polymer science, and other disciplines.

- From 2000 to 2010, grants to study bio-inspired design tripled, and the value of grants quadrupled to $93 million. The worldwide economic crisis reduced grants across the board since 2008, but 2011 saw a sharp increase in both grant numbers and total value.

The Da Vinci Index stands to have a positive impact on awareness, policy, and the business uptake of biomimicry. At this point, however, the index doesn't yet report on revenues generated by biomimetic products and companies. FBEI acknowledges that privately held companies are reluctant to disclose their financial data. My team at PAX was interviewed for the economic impact report and this was true for us. You'll see that fellow biomimics, who I interviewed for this book, have the same caution. I guess from a biomimetic standpoint, it makes sense. Squirrels don't tell where they hide their nuts. Public corporations may show more financial data to the public, but they rarely break out revenue numbers for individual products. It may take some time for biomimicry to gain enough history that companies are more willing to share financial data. Given the trends seen by FBEI and other research institutions, that time is rapidly approaching.

ROOTED IN HISTORY

It's not hard to imagine the early evolution of rafts and canoes from the sight of fallen logs drifting down a river with birds or animals perched on top. There couldn't be a better teacher in the art of pottery making than the potter wasp—all two hundred genera of them. Their techniques and designs are virtually identical to early human pottery creations. The eleven hundred species of paper wasps that mix spit with fibers from dead wood likely inspired early humans in the art of papermaking.

Potter wasp nest   Paper wasp nest

An explosion of human technological solutions appears to have occurred between 4,000 and 3,000 BCE. A visit to the Cairo museum reveals that, by four thousand years ago, ancient Egyptians had already developed many of today's technologies. From highly evolved, ocean-capable boat designs like Khufu's 143-feet-long Sun Ship, to advanced fishing nets and hooks, from pottery and papermaking to beer brewing, from bagpipes to lyres, from high-precision stone cutting to pyramid construction and plywood manufacture, there are literally hundreds of sophisticated inventions. Many echo the form and function of local animals and plants. Architectural columns clearly replicate the structure of lotus stems. A piece of armor found in a pharaoh's tomb looks just like metal fish scales sewn onto a fabric backing. Even the proportions of Egyptian tombs matched the growth ratios of trees.

Let's jump forward to the ancient Greeks—the founders of modern math, geometry, and science. Plato, Pythagoras, Aristotle, Heracleitus, and Socrates were all students of nature and sought to explain the nature of existence and matter. Pythagoras, in particular, was fascinated by the geometric proportions found throughout the natural world. Before Pythagoras, there is little evidence that musicians tuned their instruments using any particular system or scales. It's understood that Pythagoras experimented with a monochord, a single-stringed instrument with a moving bridge, to identify the way that plucking a string of various lengths creates particular musical notes. The proportions that he identified to be most harmonious happened to match the proportions of animal and plant growth (which we'll investigate later in this book). His observations were the foundation of the Western scale of music. This is a great example of isolating natural elements and combining them into a new art form. So magical seeming were his discoveries to conservative authorities that he feared for his life and started a secret society to study nature's mysteries more deeply. Heraclitus theorized, by observing the natural world, that everything in existence was created from flow in nature—which physicists now agree to be true. Meanwhile, Plato saw particular angles and proportions everywhere and developed the science of geometry. More than two thousand years later, Einstein echoed Plato's understanding that "God ever geometrizes."

A great leap in biomimetic, European architecture occurred with the return of the Knights Templar from the Holy Land Crusades. It appears that they brought back an understanding of natural geometric proportions that classical builders had used to construct the noble buildings of the ancient world—Solomon's temple, the Pyramids, and the Parthenon, to name a few. These design processes had been lost to Europe during the Dark Ages but were kept alive by the Islamic world. Following the Templars' return to Europe, an explosion of "golden," or "sacred geometry," cathedral building took place. Masterpieces like Canterbury, Chartres, Notre Dame, Amiens, and others were constructed to conform to nature's underlying proportions.

The Renaissance saw a spectacular resurgence of biomimicry in art and architecture. The great artists knew that art must echo the mathematical ratios of life to look realistic. By understanding the very precise way our eyes read distance, depth, and light, for example, they replicated our three-dimensional world on two-dimensional flat canvases, by painting objects in the front of a canvas that were larger than those at the back. Raphael, Michelangelo, Botticelli, Donatello, Titian, and many others worked intensively to copy nature's proportions in their sculptures, paintings, and architecture.

Raphael's *The Marriage of the Virgin*

Leonardo da Vinci was perhaps the greatest biomimic of all time. Not only did he precisely adhere to nature's proportions in his art but also spent the last ten years of his life studying—even obsessing over—the geometry and motions of natural flow. Based on years of observing birds in flight, Leonardo, the world's first fluid dynamicist, designed flying wings, a helicopter, and numerous other machines. His understanding of how the human heart actually operates—through manipulation of whirlpools—has only been rediscovered in the past decade. Even five hundred years later, the depth of Leonardo's insights into the secrets of nature's form and function cause scientists to marvel.

Not just a series of geniuses understood the merits of natural solutions. In past centuries, biomimicry underlay entire industries and supported generations of craftsmen. For example, consider how world history has been shaped by the winners of armed conflicts—who were often victorious because of their superior armor. Animals from armadillos to fish, lizards, and snakes all benefit from nature's highly evolved, flexible, scale armor. Lightweight, with a wide range of movement made possible by their overlapping design, scales have been studiously copied to protect warriors and their horses for millennia. Soldiers found that scale armor was cheaper, lighter, and more effective against attacks than other forms of armor—including chain mail. First evidenced in ancient Egypt and China, scale armor was widely used by Mongolians, Tibetans, Japanese, Russians, Indians, Koreans, Persians, and throughout the Byzantine Empire. Scales were crafted from metals, horns, leather, and cuir-bouilli—leather boiled in a mixture of water, oil, and urine to make it tougher. Roman centurions wore exceptionally effective *lorica squamata*—scales hammered out of bronze or iron. In all examples, the similarity to reptilian scales is obvious. Coincidentally, one of the latest evolutions in body armor is an adaptation of the molecular structure of ceramic that is benefiting from some of the same principles. Invented by Murray Neal for Pinnacle Armor, Dragon Skin is the first flexible armor that is marketed as being able to defeat rifle rounds.

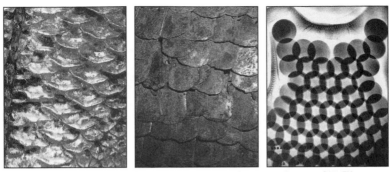

From left to right: fish scales, Roman scale armor, Dragon Skin™

Another bio-inspired invention that has had a profound impact on the course of human history is barbed wire. Historians now state that barbed wire has had as much importance in the settlement of the American West, Australia, and other large land areas as guns, the railroads, and the telegraph. Invented by Michael Kelly in 1868 and perfected by Joseph Glidden in 1874, barbed wire fencing could be rapidly erected over long distances and corral even the wildest of cattle. Originally marketed as "thorny fence," barbed wire is a simple iron mimic of an even older, biomimetic fencing woven from naturally occurring briars. As you can imagine, the impact of barbed wire on migratory animals and indigenous, nomadic peoples such as Australian Aboriginals, who had wandered freely for fifty thousand years, or Native Americans, who called it "devil's rope," was devastating.

The leafy sea dragon's appendages mimic seaweed while the
flounder is able to change its color to disappear into its surroundings.

Nature is also the true master of camouflage. From the almost invisible stone flounder and octopus to the snow fox, leafy sea dragon, and praying mantis, stealth is the difference between life and death. Unless mating, nature's creatures do not like to be seen or heard, because to be seen often means to be either eaten or starve. This hasn't been lost on warriors. Native peoples around the world adorned themselves with feathers, branches, animal skins, and paints to disappear on the hunt; and all modern armies have studied nature's systems and adopted her finer points. Nature uses several types of camouflage including: mimetic, which works by mimicking an object's shape or color, like a leafy sea dragon, octopus, or stick insect; or disruptive, which uses patterns and colors like a flounder that disappears into the sea floor or a leopard in a sun-dappled forest glade. The world's military forces are great users of DPM—disruptive pattern material. So effective are today's strategies that the Australian Army recently misplaced a camouflaged truck while on off-road exercises. Authorities had to appeal through the local media to enlist the public's assistance in locating the lost equipment.

Another biomimetic invention helped to revolutionize one of the world's most important industries: timber cutting. At the turn of the twentieth century, foresters began developing gas-powered saw blades to reduce the intense manual labor required to chop down trees with axes and hand saws. They were an improvement but required frequent and time-consuming maintenance and filing of cutting teeth. In the 1940s, one lumberjack and would-be inventor, Joseph Cox, thought for months about how he might improve the powered saws he used as a logger. One day, as he chopped firewood, his attention was caught by the finger-sized larva of a timber beetle burrowing through a nearby tree stump. Though the beetle and every part of it could easily be crushed by Joe's firewood ax, it was steadily chomping its way through hard timber, regardless of the direction of the grain. Joe studied the larva's alternating, sideways-sweeping jaw design and went to work to replicate it in an improved saw chain. First sold in 1947, his saw chain was so popular that by 1951, Joe's company was already earning more than $1 million in revenue. The basic design of that original chain is still used today, under the leading global brand, Oregon. Revenues for Oregon saw chains have topped $300 million.

Zoopharmacognosy is the long-winded scientific label for studying animal self-medication. You may have seen your pet cat or dog chewing on grass when it's unwell. Chemicals in animal-chosen medicinal plants have been shown to have antibacterial, antiviral, antifungal, and antihelminthic (antiparasitic worm) properties. Wild chimps eat *Vernonia amygdalina* to rid themselves of intestinal parasites and aspilla leaves for rheumatism, viruses, and fungal infections. Other animals chew on charcoal and clay to neutralize food toxins and rub themselves with citrus, clematis, and piper for skin ailments. Pregnant elephants have been seen to walk miles to find a certain tree of the Boraginaceae family that brings on labor. There are undoubtedly many more remarkable opportunities to be understood and adapted.

Pharmaceuticals are essentially biomimetic in principle, but are not often designed to have no side effects. Drugs were historically created from natural substances; the word *drug* comes from the Dutch *droog*, meaning "dried plant." As evidenced in Neanderthal archaeological digs, natural medicines have been in use for more than sixty thousand years. Excavations have revealed the use of at least seven herbal remedies that still show proven therapeutic value, including ephedra (as a cold remedy), hollyhock (poor man's aspirin), and yarrow (wound dressing). It's thought that early humans may have learned about some of these cures by observing sick animals seeking out plants that were rich in beneficial substances.

More than seven thousand compounds used in modern medicine have already been derived from plants, with thousands more waiting to be identified. In fact, of all the pharmaceuticals marketed in the United States today, more than 50 percent are synthesized from isolated active ingredients in medicinal plants—from stimulants to sedatives, painkillers to chemotherapy, detoxicants to antioxidants, and cardiotonics to antidepressants. As we know, often these synthesized compounds create side effects, and sometimes even very serious unwanted consequences. Though we do see headlines about overdose deaths from illicit drugs, there are actually relatively few of them compared to the more than three hundred thousand serious complications each year, just in the United States, from the improper use of, or interactions between, prescribed pharmaceuticals. In fact, a report in the *Journal of the American Medical Association* estimates that there are many more deaths from adverse drug reactions than from road accidents—approximately one hundred thousand per year. Interestingly, there are almost no major complications or deaths reported from cultures that still use original medicinal plant compounds, which often do at least as effective a job as their pharmaceutical equivalents. This is fortunate, when you consider that 80 percent of the earth's people can't afford or don't have access to pharmaceuticals and rely on plant remedies.

The herb ephedra has been used in China and India for five thousand years as a stimulant for cold and flu sufferers. Later known as Mormon tea, ephedra is now synthesized as pseudoephedrine and is found in many marketed cold remedies. (Unfortunately, it's also a key ingredient in the illicit manufacture of highly addictive and destructive methamphetamine.) Quinine, from the bark of the rain forest tree, *Cinchona ledgeriana*, is an effective preventative to malaria, one of the greatest killers of humanity, with up to one million deaths per year. The heart drug, digoxin, is synthesized from the foxglove flower. Aspirin's principle ingredients were recognized in willow bark by Hippocrates around 400 BCE. It was named and marketed by Bayer in 1899 and is still one of the biggest selling drugs in the world.

The pharma-biomimicry list is extensive. The indispensable bioinspired painkillers morphine and codeine are derived from opium. In the 1890s, Bayer developed a cough suppressant from synthesized morphine that it trademarked Heroin. With the United Nations estimating that there are currently between twelve and twenty-one million regular users of heroin worldwide, heroin could be considered the bestselling drug brand of all time. Even the 1960s drug LSD is synthesized from a natural rye fungus called ergot. Incidentally, it's been proposed by some historians that the European witch hunts may have been a result of so-called witches hallucinating after eating ergot-molded rye. The theory is that their antics were caused by inadvertent "bad trips," which resulted in these unfortunate wretches being branded as witches, with up to one hundred thousand burned to death. Some might argue that heroin and LSD are examples of biomimicry gone wrong, but I believe the fault is with humans choosing to synthesize and distribute these molecules without consideration and management of their consequences.

As a subset of pharmacology, how could we estimate the value of antibiotics to humanity? Prior to World War II, even a scratch or ingrown toenail could turn septic and kill you. Diseases like tuberculosis, pneumonia, syphilis and gonorrhea, plague, even a tooth abscess, and a hundred other ailments were often death sentences. In one of the most fortunate accidents of all time, Alexander Fleming noticed that a petri dish full of virulent staphylococci bacteria he had cultivated for research had been contaminated by mold. He was about to wash out the dish, when he realized that the mold had killed some of the bacteria. The rest of this biomimicry story is history. Named by *Time* magazine as one of the hundred most important people in the twentieth century, Fleming originally called his antibiotic "mold juice." Today, more than seven million pounds per year of antibiotics are administered to ailing patients and countless lives have been saved. Although Fleming became aware of the real risk of emerging antibiotic resistance and lectured widely on the threat, he was largely ignored. Despite Fleming's cautions, overuse of antibiotics has resulted in the evolution of superbugs—bacteria that are highly resistant to antibiotics. The Centers for Disease Control waited from World War II, when data began accumulating, until 2008 to release its first report on the side effects of antibiotic use and overuse. Now considerably more people die each year from antibiotic-resistant staphylococcus than from AIDS. Fortunately, new biomimicry research is under way that could result in whole new classes of bacteria-suppressing drugs that don't result in resistance or distressing side effects. We'll hear more about these later in the book.

Future profits come to those who anticipate trends and work toward meeting them with services or products. In this case, the trend is clear. Biomimicry has always been a great source of wealth and opportunity. Now it can be a far greater wealth generator and problem solver than ever before. Most opportunities are still waiting to be identified or marketed, so the potential for intellectual property creation, new manufacturing methods, and breakthrough chemicals and materials is immense. Nothing short of the overhaul of the entire industrial sector is possible.

FROM THE BEACH TO THE BOARDROOM

Although I didn't sense biomimicry's potential to revolutionize the world's economy at the time, I started my career in biomimicry when I was ten years old and fell in love with the beach. We lived a block away from the warm waters of the Indian Ocean. On long, hot, Australian days, I pedaled my bike home from school as fast as I could, picked up my mask and snorkel, and rode to the pristine shore of Cottesloe Beach. I collected seashells, hunted for octopus in rock pools, gathered abalone, dived for lobster, and spear fished every chance I had. I daydreamed about swimming and fishing, surfing and snorkeling pretty much all day at school and when I lay in bed at night. Being in the ocean was my delight and joy—powerful motivation for a career choice.

I was fascinated by how easily, gracefully, and efficiently fish could propel themselves through the water—much faster than my ungainly struggles to catch up with them. I noticed that the wavelike motion of their fins and bodies was similar to that of eels, snakes, and the twisting patterns of seaweeds. Seaweeds can be quite fragile if you pull on them. I discovered this when I caught hold of their stalks while snorkeling near reefs or rocks. As a surge from a wave washed me away from where I wanted to be, I grabbed on to seaweeds to anchor myself. Often they broke. Yet even in the wildest storms, these same weeds were able to stay intact and survive. The surge of huge waves couldn't break or dislodge most plants from their grip on the ocean floor. It became apparent to me that they were adapting their shapes to the path of least resistance to relieve the onrush of water. Although at first it appeared that the weed fronds were moving chaotically, long observation showed me that all the plants were generally bending to a particular swirling pathway.

Around this time, my education was placed in the hands of the Jesuits, the intelligentsia of the Roman Catholic Church. My schooling focused entirely on the classics and prayer. Mass and benediction, daily and weekly events, were compulsory. The school chapel was an austere building. Plain sand and clay bricks with confessionals on the left side and Stations of the Cross mounted on the walls above them. The right-hand wall was mainly glass and looked out onto a close-cut lawn and a hedge of dark green hibiscus bushes. Not much going on there, either. Bored out of my wits, I examined every nook and cranny, counted every pattern on the ceiling tiles and even the number of stitches on the collar of the boy in front of me in the next pew. I practiced hyperventilating to improve my snorkeling skills, until one day I blacked out as I stood up from a kneeling position. I crashed backward onto the pew, knocking it over, causing uproar with the boys, and interrupting the service.

Half nautilus shell

Bishop's crook

In one of these deadening sessions during my tenth year of life, the archbishop turned up to say Mass. He entered the church, elaborately clothed in a white and gold dress, caped, and carrying a long stick with a spiral on top. The spiral caught my eye; it had the same curve I had seen in many other forms at that school and matched the shapes of seashells I'd collected at the beach. Identical spiral patterns adorned the bound covers of the missal and the Bible, the tabernacle, the altar cloths, and were even carved into the pedestals upon which stood the stiff white statues of Jesus, Mary, and Joseph. The frames of the Stations of the Cross also had spirals in each corner. I began asking about this shape. Although no priest was able to answer my questions as to its origin or purpose, I sensed that the spirals I saw in church somehow connected and pointed to natural forms.

With the beach a part of most of my days, I survived my high school years and made my way toward a career in electronic engineering. Unfortunately, being red-green color-blind in the days of dozens of colored, twisted-strand wiring, it was not a very satisfying experience. I lasted two years before quitting to join the Department of Fisheries and Wildlife, where I was paid to be on and in the ocean, protecting the fish and animals of Australia—a job that suited me very well. I worked my way up the ladder, becoming a patrol vessel captain and eventually a senior field officer. Along the way, I started a cooperative for local fishermen, invented some measuring devices for the fisheries industry, and had many adventures in the wild.

By 1980, I was frustrated by the increasing trend of real estate developers taking over wildlife refuges to build housing projects. Usually the complex would be named after some unique quality of the area, never mind that the project had wiped it out. "Quiet Waters," "Pelicans' Nest," "Dolphin Bay." . . . I would work for years to protect a wetland or a nesting site and then watch as bulldozers went in to strip the land and build homes for an increasingly affluent Australian population. I had seen enough. It struck me that speculators and bean counters were running the show. If I wanted to protect nature, I needed to find a way to show them that it could be more profitable to conserve nature and study its many wonders than to chop it up. I left government service and enrolled at university, to study economics and then psychology and comparative religion.

"The rise in share price is showing no sign of slowing down," my secretary, Avi, said as she walked into my office.

It was February 1985. As founder and CEO, I had placed ERG Australia, Ltd. on the stock exchange the previous month. Described as one of the most successful initial public offerings of an industrial company in Australian history, the share price had already increased nearly 500 percent. Quite strange, really; I was having trouble getting used to these events. I had founded the company almost two years to the day before its launch on the stock exchange. After university, I had traveled the world to study with religious leaders and mystics, and returned to Australia penniless. There was an economic recession and jobs were scarce. My previous career as a Fisheries and Wildlife officer hadn't equipped me with specific skills to make a living outside government service, but ever the optimist, I decided to launch a company that would study and apply the energy-saving strategies that I'd observed in nature. Energy Research Group (ERG) seemed like a fitting title that occurred to me one night at the dinner table. The next day I registered the name.

I was living in a bedroom graciously provided by a friend. The same friend also loaned me $20 to buy a new white shirt so I could better look like a business executive. The shirt, determination, my experiences with starting a fishermen's cooperative, and a childhood spent trading scrounged scrap metal and soft-drink bottles was all I had to launch me into the heady world of high-tech start-ups. I soon found two partners, both engineers, with a good idea for an electronics product that looked like it could make money faster than my nature-based work. I decided to suspend my agenda and come back to biomimicry (though the term hadn't been coined yet) when I had the capital to support it. There were, of course, the usual naysayers—some quite hostile to our ambitions— and any step I took was into new territory for me. However, I was not blessed or burdened by much business knowledge or experience. A series of fortunate meetings, a great deal of hard work, and two years later, we were "an overnight success."

The following years were tumultuous. A few weeks after our debut on the stock market, I was in London negotiating a license for our products. ERG had excellent prospects, a large pot of cash, skyrocketing share values, and I had just completed a major deal with Australia's largest outdoor media company involving a very significant up-front fee. While I was away, a group of industrialists and financiers, including some of our newly appointed ERG board members, succeeded in a carefully planned hostile takeover. They fired me and more than half the staff for fear that we wouldn't be aligned with their conquest. I was stunned.

In its own way, life in the corporate jungle is biomimetic. Nature has its means for finding balance overall, but it has no word for justice at the individual level. Some birds specialize in stealing the catch of others. Claim jumpers in any gold rush do the same. Aggressive companies often beat, and sometimes eat, their competitors.

After selling my ERG stock, I spent the next five years traveling and starting a number of new ventures in England and Australia. I learned a great deal from each of those businesses, but none of them satisfied my motivation to show industry that there was more profit to be made by learning from nature than from stripping it.

One day I found myself walking on a pristine beach in Australia, my feet sinking into the sand as rushes of clear Indian Ocean waves washed over them. I stopped and breathed in the salt air, polished clean by its travels across thousands of miles of unpolluted ocean. The sun on my shirtless back reminded me of my childhood days spent exploring beaches and tide pools. As a young man, it was that fascination which led me to become a naturalist and join the Department of Fisheries and Wildlife.

I remembered an obscure career that I'd learned about when, as part of my conservation duties, I'd served as the pearling superintendent of Western Australia. Beachcombing was still an official occupation under the old government Pearling Act. Enacted in the late 1800s and still in force, though many years out of date, a beachcombing license was used by people who collected washed-up pearl shells from the tropical beaches of Australia. Pearl shells were a valuable, highly regulated commodity in that era. What a great way to make a living, I thought. What could be better than beachcombing? But what could I find on a beach that would be valuable in today's world? Pearl shell wasn't worth as much since the arrival of plastic buttons. Maybe I could collect wild, edible seaweeds from the reefs I was so familiar with from years of diving and spear fishing. What else was possible?

I turned my gaze from the ocean and was about to head home, when I noticed a broken seashell on the sand. All that remained was the graceful, spiraling center. I remembered my childhood fascination with swirling seaweed letting the force of waves go past and the similar shape of whirlpools above the bathtub drain. Here in front of me, was the same spiraling pattern. I picked it up and turned it over in my hand. I knew that the snail that had once lived in the shell was like a liquid as he grew. I also knew from my years of field research that nature always uses the least energy and least materials to get the job done. The shell's creator built the path of least resistance, friction, effort, and materials into his home.

My "eureka" moment

In that instant I knew how to make a living from beachcombing. All I had to do was study the optimized shapes common throughout nature and adapt them to the design of all manner of equipment. In particular, I felt the swirling shapes that I had observed time and again could be applied to industries and products that interacted with liquid, gas, or heat flow, and that's just about every industry and product in the world. It proved to be a thrilling, engrossing, and challenging process.

The study and application of this observation became my life's work for the next twenty years—and made me a comfortable living from doing what I love and for which I feel a strong sense of purpose—the betterment of humanity and the protection of natural habitats. The simple act of deciding to find and follow what fascinated me—beaches and beachcombing—launched me into not one but two richly satisfying careers: naturalist and biomimic.

Chapter 2

GOING WITH THE FLOW

The Shape of Things to Come

At the dawn of thought, Mr. and Mrs. Cavesperson saw mystery all around them. Chaos, randomness, and inconsistency were everywhere. Not a single straight line existed on earth. No wheels, fasteners, tools, or fences. Just virgin wilderness with wild streams and patches of grassland. Humans were not as high up on the food chain as we are now, and had very little ability to manipulate their environment and develop the relatively modern feelings of invincibility and control over life. Birth and death were completely unfathomable and absolute secrets—which, essentially, they still are. But wait. In the midst of wild disorder, a recurring pattern is observed, an aesthetically captivating and evocative pattern, and it's everywhere around them. In the curving horns of their prey, the tendrils on vines dripping with grapes and berries; swirls in seashells; eagles' talons; chameleons' tails; flower petals, seeds, and ferns; octopus's tentacles; whirlpools and dust devils; and snakes. The spirals are all similar to one another and all very different and distinct from the usual, disarrayed jumble. What did they think of all this? How did they explain this uniformity?

We do know that the earliest humans drew these shapes on cave walls. The first shamanic drawings were of stick people and animals. The next type of drawings included spirals. From this chronology, one might surmise that humans first drew "self," when they recognized self, and then drew spirals when they recognized underlying order in the chaos. It is quite remarkable that virtually all the world's peoples and cultures—

even those communities living on different continents, with no possibility of communicating with one another—developed an appreciation and reverence for this pattern. Indeed, many of the world's early religions' most powerful spirit figures or icons were of a spiraling serpent—even in parts of the world where no snakes existed.

The spiral grew in importance and was carved into stone temples and burial chambers of the megalithic period, beginning some seven thousand years ago. By the fifth millennium BCE, spirals were tattooed on faces in China and Europe, painted on clay vessels, or woven into grass baskets and clothing in every inhabited country. The mathematics underlying nature's spiral had been worked out and incorporated into architecture. The Minoan, Babylonian, Greek, Persian, Roman, and Chinese civilizations all used the proportions of natural spirals extensively for aesthetic purposes. It became a most prevalent symbol for the spiritual, God and the deeper meanings of life and life force in almost all religious movements. Serpents, dragons, and other expressions of single and double spirals, such as the yin-yang symbol and the hermetic caduceus, proliferated.

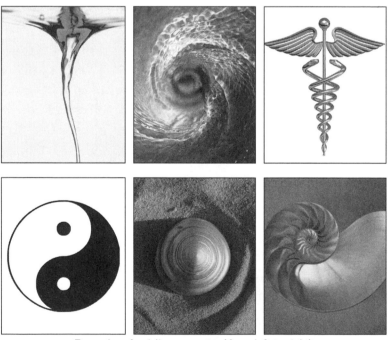

Examples of swirling geometry (*from left to right*):
whirlpool side view, whirlpool top view, caduceus,
Yin-Yang, seashell top view, half nautilus shell

Humans have long seen the swirling geometry of a whirlpool or a tornado reflected in the growth of seashells, whether viewed from the top or the side.

In 1638, French philosopher and mathematician René Descartes dubbed the snail shell spiral "the equiangular spiral," when he noticed that lines drawn from the center of the spiral intersected with each spiraling arm at identical angles. If we measure the length of each piece of the line that intersects the spiral, we find that each one relates to the next by the same proportion. Equiangular spirals that exhibit the growth factor of 1:1.618 are widely known as the golden ratio, or divine proportion spirals. This same ratio can be seen throughout living bodies, from the segments of our fingers to the curls of our hair and eyelashes.

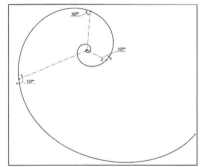
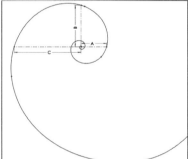
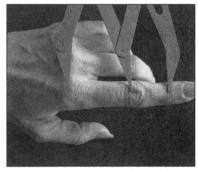

From left to right: equiangular spiral, golden proportion, golden proportion in finger segments

The equiangular spiral is the only three-dimensional, mathematical curve that retains the same shape while growing only at one end—handy for a growing organism. A snail can't expand the inflexible walls of its shell; it can only add on to the open end. The shape of its chamber does not change as the snail grows but continues to match the contours of the animal that inhabits it. If you magnified the spiral of a small snail shell, it would be identical to the spiral of a large snail shell, a property called self-similarity. This quality emerges at all scales of size, just as we see in fractal geometry. The fractal, an important tool in mathematics, is a term coined by Benoit Mandelbrot in 1975 for a human-generated, self-similar, often spiraling, multipart geometric shape, wherein each part is a reduced-size copy of the whole.

Computer-generated fractal art, representing
mathematical algorithms, which bears a striking
resemblance to romanesco broccoli and other natural shapes

We see similar shapes throughout the natural world. A mollusk seeks the path of least resistance as it grows and slides in and out of its shell. Likewise, the hermit crab that moves into the shell after the mollusk departs, grows himself into the same shape in order to facilitate coming and going, when he's not hiding in it or carrying it around for protection. The coconut crab is an impressively large, coconut palm–climbing hermit crab that tears coconuts apart with its claws and no longer finds wearing a heavy, hand-me-down shell useful. Any arthropod capable of ripping a coconut to pieces through crushing brute force probably isn't too worried about protecting its butt with a cumbersome hard hat. Its body still grows into a three dimensional spiral, without ever using a shell.

Turtles are remarkable animals that fully exploit spiral geometry. Their shells and heads benefit from the same strength-to-weight ratios as birds' eggs. The turtle is slow on land, but watch him go in the ocean—astounding performance. Turtle shells are superbly streamlined and provide excellent lift for their flight through water. Their immensely strong beaks are built to spiral geometry, as are their flippers. A turtle's front flippers don't actually flap but rotate as they propel the animal forward—like a sculling oarsman on a boat. This rotation, like a bird's wing, rolls up the animal's wake energy into spiraling ring vortices that it then recycles and leverages off to increase thrust. Now, with rapid expansion of computer tools and high-speed photography, fluid dynamicists can learn to adapt and apply these extraordinarily efficient strategies to the design of cars, boats, and planes.

Spiral shapes weave through everything. Magnetic fields are arranged in spirals. Mushroom spores propagate in spirals. And no matter what our race or religion, size or shape, we humans are made of the same all-pervading spiral geometry. This is very apparent in the swirling shape of heart muscles and skin pores.

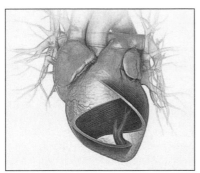

From left to right: spiraling heart muscle, human skin pore

The cochlea of all mammals matches the spiraling design of a seashell, while the shape of our outer ears echoes the curled-up embryos of humans and many other animals—a feature utilized by acupuncturists who treat ailments in various body parts by pinning needles into the location on the ear that corresponds to those body parts.

From left to right: tun shell, human cochlea

Like the Renaissance masters did in their exquisitely realistic paintings of human faces, a cosmetic surgeon seeking to reconstruct a severely injured face gets by far the most realistic results if she applies the same mathematical ratio to the size of facial features and the distance between them. Even the size of our teeth and the way they relate to one another is based on the same ratio, which helps dentists create natural-looking dentures.

Within the structure of our own chromosomes, DNA is described as a double helix. The image is of two parallel, curving lines, wrapping like a twisted ladder. How can you twist a ladder? The sides must alternately curve in and out. At the other end of the continuum, Einstein's curved space doesn't result in parallel lines either. If space is curved, then all movement must always either be converging or diverging. Buckminster Fuller wrote about DNA as a nonparallel-sided, dual spiral that has an unzipping angle; and the scientific artist Kenneth Eward has recently depicted a DNA molecule as it actually is. His representation shows a clear similarity to the geometry of a seashell.

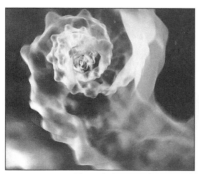 

From left to right: artist's rendition of DNA, seashell

Any close, in-depth look at the nature and prevalence of the equiangular spiral invokes awe at the beauty and simplicity of this universal blueprint. From the movement of the smallest particles observed by science to the largest structures in our universe to solar flares traveling at up to one third of light speed to the formation of clouds behind an island and weather fronts on the TV news to the growth patterns of all living things, the list appears to be endless.

The proportions and mathematics of nature's spiral were used in many icons of pre–twentieth-century architecture, including the Pyramids, Parthenon, Taj Mahal, Angkor Wat, Chartres Cathedral, the White House, and many other of the greatest and most treasured buildings of all time. Who doesn't have a sense of well-being and awe when in, or seeing, one of these magnificent buildings? The interiors of medieval cathedrals in particular evoke a powerful sense of stillness, reverence, and a quiet mind.

The most difficult of all the sciences is often considered to be fluid dynamics—the study of gases and liquids in motion. Swirling smoke from a fire or roiling banks of fog or clouds present huge challenges for scientists. This motion is so complex that it is extremely difficult and unreliable to predict, resulting in varied and often erratic weather forecasts and inaccurate plotting of the course of hurricanes. You might ask what relevance the science of fluid dynamics has to you and me. Since the start of the industrial revolution, the human family has burned over 350 tons of fossil fuel and produced a trillion tons of carbon dioxide (one ton of carbon produces 3.67 tons of carbon dioxide). More than half has been burned in just the past fifty years, and the rate is rapidly increasing.

What did we use all this energy for? We used most of it to overcome friction, drag, and gravity. If there were no such thing as friction or drag (resistance to movement), we would use a small fraction of the energy we currently generate each year. Your car, traveling down the freeway, is struggling against the obstacle of air. Just stick your hand out the window to see how powerful this obstacle is (though do it safely). Everything moving through a liquid or gas (whether car, plane, or bullet) and every device where liquid or gas moves through it (like a pipe, chimney, or pump) suffers from this friction or drag. That means just about every device or machine made by humans. Fluid dynamics tries to understand and limit this drag to reduce energy waste and increase performance, which is why studying nature's spiral shapes and structures presents such an important line of inquiry for the future of our planet.

There is not a scientist on earth today who can fully explain and replicate the flight efficiency of a single insect or the swimming ability of any fish. No man-made pump and piping can match the efficiency of the human heart and vascular system. Globally, we devote huge amounts of energy to cool our computers, yet nature's super-computer, the human brain, has no heating problem at all. As we'll see repeatedly throughout this book, nature always uses the least energy and the least materials for the job. Nature and humans use dramatically opposed strategies for drag reduction and neither borrows from the other.

What is nature's strategy? To answer, let me pose another question: What is the shortest distance between two points? Say, between one set of goalposts on a football field and the goalposts at the other end? The answer is obvious: a straight line. And what is the most efficient way to move fluids or energy or an object through fluids between two points? Is it still a straight line?

Almost all engineers and classically trained scientists would say yes. Yet after billions of years of evolution, nature invariably transfers fluids or energy or an object through fluids from point A to point B not in a straight line but in the same spirals that we see in the whirlpool above our bathtub drains. Engineers and physicists call it turbulence. While engineers have covered the earth with straight pipes, chimneys, and drainage canals, and straight, square buildings, nature never, ever, uses a straight line for anything or any purpose. The archaeological record since the dawn of time fails to produce a single example. All life, and even crystals, form from a liquid state and therefore are born of nature's spiraling flow geometry. Even the facets of diamonds are not straight when looked at under a scanning electron microscope.

From left to right: electron microscope photo of
diamond surface, as seen by the human eye

Lift one hand in front of you now and inscribe a circle in the air with your finger, taking one second to complete it. As you're completing this process, the earth that you're sitting on is spinning on its axis and progressing radially with the rest of the solar system through space at the same time. Our earth travels through the solar system at 18.5 miles per second, and the entire solar system is barreling through space at 155 miles per second. So by the time your finger comes back to the starting place of your air circle, you will have arced more than 155 miles through space. Your circle looks like an expanded, uncoiling spring that is more than 155 miles long. In the same way, a brick falling from a tall building seems to travel in a straight line, but in the seconds that it takes to hit the ground, it has actually traced a long spiral relative to the universe. This applies to linear accelerators or anything traveling in what seems to be a totally straight or planar line. Incidentally, if a person is lost in a featureless desert, it has been found that he or she doesn't actually walk in circles as popularly thought. In reality, the meanderings follow spirals.

The geometry of this spiral gives us clues to all other paths of movement in our universe. Consider our own solar system with the planetary paths drawn out over time, as seen in an artist's rendition.

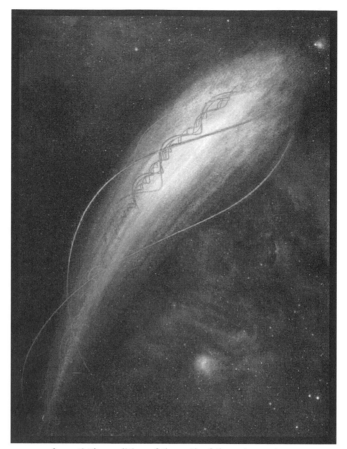

An artist's rendition of the path of the solar system

In 1609, Kepler proved that planets move in ellipses—or oval shapes—which is still accepted. But this is not truly correct, since an ellipse meets itself again. If you draw an eclipse on a sheet of paper, to close the diagram your pen must return to where it started, like you originally imagined the air circle, above. But all planetary bodies move in ever-increasing or decreasing spirals through space and never retrace their steps, once again, like an uncoiling spring. This more precise description opens doors to valuable inquiry.

If we create a large flat surface of concrete and leave it in the bottom of a riverbed for a million years or so, it will end up looking like the swirling rock canyons of the southwestern United States.

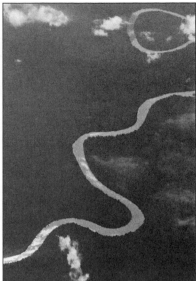

From left to right: erosion patterns, meandering river

We see streams and rivers the world over (and evidence of ancient ones on Mars) all having similar meandering swirling patterns, never straight lines. In fact, wherever humans create straight flow paths such as stents in arteries; viaducts taking storm water to the sea; or pipes in any air, water, or oil supply, we see that the fluids always twist and turn and leave deposits in spiraling shapes regardless of all attempts to keep them flowing straight.

It would make sense that if straight was best, nature would have evolved its designs that way—at least once. But there are no straight sides on a bumblebee, a fish, a plant, or an artery. However, nature isn't rigid about what we might call empirical perfection in its spirals. Environmental vagaries and fluctuations lean against and distort the pure form as it is manifesting. A tree might survive droughts, fires, high winds, and insect attacks, each causing aberrations in growth patterns. A seashell-dwelling animal may be in rough or calm waters, with optimum nutrition for laying down its shell, or not. Yet even with these distorting challenges, from the smallest to the largest movements in our universe, nature constantly converges on the same, self-similar, spiraling geometries, even if species or continents or billions of years and light-years apart.

Clockwise from top right: butterfly tongue, hair growth pattern, ammonite fossil, atomic particle decay pathways, spiral galaxy, aloe vera, hurricane, spiral horns

PAX

Archimedes of ancient Greece was among the first recorded biomimics. One of his greatest inventions we now call the Archimedes' screw, a device that's still used in the present day and, in one form or another, underlies many of the fan, pump, propeller, mixer, and turbine designs of the twenty-first century. One can imagine Archimedes' aha! moment when walking through the woods, observing rising mist and water flowing in streams. He must have realized that nature always moves water and air in swirls and spirals. He likely surmised that he, too, could use a spiral to do so. This was a profound insight. His device worked—it pumped water. It still works—in countless millions of machines across the world. But Archimedes' curved-screw pump had limitations—with two-dimensional, parallel sides that only approximated the multicurved spirals that we see throughout the natural world. Was his interpretation the optimum way to do things?

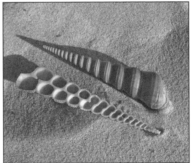

From left to right: Archimedes screw pump, nature's logarithmic spiral

This is one of the many questions I face at my biomimetic company, PAX Scientific, Inc. Nature has evolved to accomplish the greatest effect with the greatest efficiency. If nature could have done it best by using Archimedes' simple, two-dimensional screws, it would have; but instead, nature moves water and other fluids in three-dimensional logarithmic spirals, not the clock-spring-shaped pumps that Archimedes pioneered. I guess we need to cut Archimedes some slack, though. Building a three-dimensional, logarithmic pump or propeller can be challenging, even with today's technologies, and would have been impossible in his time.

The great names of science of the nineteenth century (the century of discovery) were convinced that the universe could be explained in terms of the geometry underlying nature's spirals and whirlpools. But we haven't applied that thinking to modern science and technology. Nature's geometry as mentor for fluid dynamics—one of the most important of all disciplines relating to human machines—has been completely ignored.

Why? The beginning of the industrial revolution saw the pressing, stamping, and rolling of flat sheets of metal and flat, sawn wood. Flat sheets could be rolled into cylinders and cones, but not nature's three-dimensional curves. There you have it. Scientific and engineering tools were developed to build and deal with the flat and the straight. Plus, we have had straight-line thinking as the cornerstone of our scientific world. As a race, we abandoned the notion of a flat earth about five hundred years ago but still think in terms of straight lines. We've become stuck in a limited paradigm. The whole known universe is made of and according to nature's spiraling geometries—and nature uses them exclusively to move energy. The entire human family before the industrial revolution venerated those shapes in art, architecture, and religion—yet we in the modern age of science have ignored it almost completely. I still find very few engineers and scientists that have even heard of this geometry, let alone investigated the complicated math, physics, and mechanical properties associated with it. What is possible if we apply nature's drag-reduction methods to science and technology?

Put another way, what sane person would work hard at her job her whole life, but every time she got her paycheck would light a match and burn two-thirds of her money? This is exactly what the world is doing day after day, year after year. Through poorly managed drag or friction, we waste two-thirds of the energy we produce and, by so doing; we're destroying our environment and atmosphere at three times the rate than if we didn't waste energy. The United States burns two billion dollars' worth of oil every day. The world burns four cubic miles of nonrenewable fossil fuels every year. That's a mound four miles long, four miles wide, and four miles high, equivalent to 21,120 feet—the highest mountain in the Andes or three-quarters the height of Mount Everest. We're very clever and resourceful at extracting and processing more and more fossil fuels but we're pouring that energy into a bucket full of holes. We're wasting a large part of this energy by trying to force flow into straight lines.

I wanted to capture the three-dimensional shape that I saw everywhere in natural fluid flow, so I could extract the essential elements and apply them elsewhere. As science had not focused on the importance of turbulence geometry and had no tools to mathematically create those shapes, I had only one option: I needed to freeze an actual whirlpool. Several years of experiments followed, sometimes messy, with a lot of failed attempts—but I was eventually able to achieve my goal.

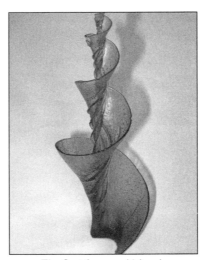

The first frozen whirlpool

I adapted my frozen whirlpool and first built a rotating mixer. For engineering applications, rotors need to be balanced to run smoothly without destructive forces and vibration. An engineer designing a mixer, fan, or propeller simply calls for two or more blades, which balance each other when the rotor is spinning. It turns out, though, that the most energy and thrust-efficient rotors have only one blade, like nature's flying maple seeds. Since nature doesn't make multibladed rotors, I didn't have an obvious champion in nature that I could find to overcome the balancing challenge. I needed to get creative.

My first mixer was two bladed. What a nightmare trying to build it. This was before the days of easy access to supercomputers and rapid prototyping technology, sometimes called 3-D printing. The shapes were already hard to build, but then I had to find the exact center at each end of my whirlpool so I could mount the mixer blades on a common central shaft. I found it almost impossible to accurately pinpoint, until I realized that whirlpools don't have a center shaft—even the central air tube that is visible in a whirlpool is a void, not a solid. My central shaft, or axle, had to be located with high precision at each end but, ideally, still be a void. Once I worked out a compromise to achieve this, I had a mixer.

I then found that if I kept growing the outer edge of the streamlines I was using to design the blades, they wrapped around each other in a surprising way. This developed into a novel pump that could efficiently move fish, rocks, broken glass, or blood cells without damage and without clogging.

PAX pumps

If I opened out the streamlines so they lay more horizontal to the center axis, it created all manner of highly effective fans and propellers—though it wasn't until we'd hired clever engineers and fluid dynamicists that we could customize our designs to our customers' performance specifications.

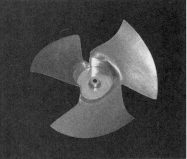

PAX fans and propellers

Cutting these streamlines in other ways, we could produce water turbine blades that don't hurt fish or clog with weeds, plastic, or rocks. These turbines didn't cavitate or churn the water or air into wasteful turbulence and could potentially simply be placed in a fast-flowing stream without the need for dams.

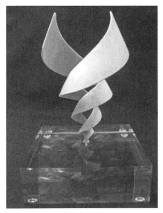

PAX turbine

One of the things I had long wanted to develop was safer, more efficient boat propulsion, particularly since a friend of mine had tragically lost his son when the boy fell overboard and was killed by the outboard propeller. Jet pumps are much safer but inefficient—if I could create a jet pump with sufficiently improved performance, boat propellers would be obsolete. Nature builds a variety of excellent jet pumps in jellyfish and the cephalopods—squid, nautilus, octopus, and cuttlefish.

I applied nature's geometries and methods to a hand-built, jet pump prototype. We tested it on a barge floating alongside our research vessel in Fremantle, Western Australia, and watched with some shock as it not only sucked mussels off the bottom of the float and spit them out the other side but also started to move us and the 150-foot boat the barge was tethered to.

An engineer friend was there for the test drive and said with some gravity, "Jay, you don't just have a better jet pump. This is a pump, a mixer, a fan, a propeller, a turbine—any kind of fluid-handling equipment."

No matter what I built—even heat exchangers, combustion chambers, pipes, fuselage designs—and although hand-built from fiberglass and far from precision optimized, all worked and achieved performance equal to or better than best in class. Essentially, we were holding the tools necessary to overhaul the entire industrial world with superior, energy-efficient technology. Clearly we were onto something very large—and somewhat daunting.

INCORPORATING INNOVATION

By the time my wife, Francesca Bertone, and I incorporated PAX Scientific in California in 1997, we had accumulated a series of patent applications, concept prototypes, and a vision for our biomimicry business. We wanted to rapidly disseminate nature's energy-efficient streamlining strategies—the result of my years of field research and experimentation on proof of concept. We had been conducting research on our marine research vessel in Australia. However, as the technology applied to so many industrial products, it was clear we couldn't possibly handle global implementation by ourselves. Besides, although a beautiful place to live, Western Australia is not a hotbed of fluid dynamics research. Our plan was to move to California and from there, license our technology to industry leaders who would deploy it through their research and development (R & D) departments. They would then pay us a small royalty for products based on our patents.

Initially working out of our home in Northern California, with a garage-based lab, I wrote a one-page letter introducing myself and what we had and posted it to the CEOs of twenty-two Fortune 500 companies. Within a couple of weeks, we had received seventeen responses, with invitations to meetings and referrals to heads of engineering departments. That level of response seemed an excellent sign. Over the next few months, I met with those CEOs or their deputies and received an enthusiastic response from almost every individual. There was also strong interest from engineers given the task of interfacing with us. However, support from their senior engineering and product development managers was less forthcoming. We learned that many of the big companies we had approached were no longer manufacturers themselves but assemblers of components or were value-added reseller companies, who put their famous names on systems that other original equipment manufacturers (OEMs) had built. That didn't daunt us, though when helpful VPs of engineering at top-of-the-food-chain companies referred us to their suppliers, we found that many had little or no R & D capacity, were unwilling to take a risk on outside ideas, or had no room in their already stripped-down budgets for innovation. Our designs found nowhere to land. It became clear that we needed to build actual products and create an apples-to-apples comparison before we could interest potential manufacturing customers.

Where to start? We created a matrix of the product areas that we believed PAX could impact and identified more than five hundred distinct market sectors—with potentially hundreds of thousands of products that we could improve. We had to focus. After analysis that included the size of the addressable market, ease of access, the cost and time it would take to develop working prototypes, the certifications and metrics of the various industries, the need for energy efficiency in the sector, and so on, we prioritized the list to fans, mixers, pumps, and propellers. We began hand-making prototypes as comparisons to existing, leading products.

By this time, we were raising working capital from angel investors. It's important to note that this was during the first half of the last decade. The tragedy of September 11, 2001, and ensuing military actions had the world's attention. *Clean tech* and *green tech* were just emerging as terms, and energy efficiency was still more of a slogan than a driver for industry. The dot-com boom had busted. We'd researched venture capital firms in the late 1990s and found only seven in the United States investing in mechanical engineering inventions. These tended to be expansion-stage investors that didn't match our phase of development. Still, we were close to the famous Silicon Valley and had a few comical conversations with venture capitalists who said they'd be interested in investing—if we could turn our technology into a website.

Instead, every six months or so, we drew up a budget for the following six months. Via a growing network of forward-thinking private investors who could see the looming need for dramatic changes in energy efficiency and the performance results of our prototypes compared to currently marketed products, we funded the next phase of research and business development.

Over the next couple of years, we built and tested a series of prototypes, started dialogues with leading manufacturers, and added business development and technical staff to our team, including mechanical and aerospace engineers. Our plan was that PAX Scientific would be an intellectual-property-creating R & D company. When we identified appropriate market sectors, we would license our patents to outside entrepreneurs or to our own, purpose-built, subsidiaries. Given my previous experience on the receiving end of hostile takeovers, we were determined to maintain control of PAX Scientific and its subsidiaries in their development stages. Creating subsidiaries that were market specific would help, since new investors could buy stock in a more narrowly focused business, without direct dilution of the parent company.

We were introduced to fellow Bay Area resident Paul Hawken. A successful entrepreneur, author, and articulate advocate for sustainability and natural capitalism, Paul understood our vision of a parent company that concentrated on research and intellectual property, while separate teams focused on product commercialization. With his own angel investment backing, Paul established a series of companies to market computer, industrial, and automotive fans. PAX assigned worldwide licenses to these companies in exchange for up-front fees and a share of revenue; Paul hired managers and set off to sell fan designs to manufacturers.

PAX fans can reduce energy requirements by up to 50 percent over competitor blades. Noise is often reduced dramatically, as well. For example, PaxFan developed a more efficient evaporator fan for domestic refrigerators. The most commonly used product was designed in 1961 and is still installed in millions of refrigerators a year in the United States. Over the decades, several attempts to improve its performance had failed. Our equivalent fan blade and motor combination gained a total energy savings of 45 percent, allowing a refrigerator manufacturer to qualify for an Energy Star rating. It also reduced noise by 40 percent. A July 2011 technical article by the American Society of Heating, Refrigerating and Air-Conditioning Engineers estimated that fans use about 23 percent of worldwide energy. PaxFan's analysis concluded that if PAX fans were fully deployed, they could save up to 4 percent of electrical energy use (about $40 billion worth annually) and subsequently up to 4 percent of energy production emissions.

As Paul and his team entered the world of fan design and licensing, we decided on a completely different business model for our next industry area. Like many biomimetic start-ups, we were outsiders to our target markets, which in our case were in traditional, rust belt industries. We had a lot of scientific data but no commercial experience in those areas, and it was increasingly obvious that market research reports and brochures distributed by manufacturers didn't tell the real story of the value chain in each industry. I recall a classic moment with a senior executive at one of the largest manufacturers in the United States. After we'd developed rapport and were discussing possible joint research, he admitted that the scientist in a white lab coat shown on their website, gazing intently at a computer screen filled with fluid dynamics images, was actually a gal borrowed from accounting. They had no one in-house who could actually produce such data.

Rather than just studying market research reports or reading trade periodicals, we learned far more about how things really worked when we contacted domain experts—career engineers and businesspeople—and asked them to teach us about their industries. This strategy has stood us in good stead ever since. Through Paul Hawken, we met an engineer with years of experience in the water industry and a masters in business administration. We supplied him with our test data on mixers and asked him to identify an emerging field where PAX's biomimetic technology could shine. Based on his research, we incorporated PAX Water Technologies, Inc. Instead of licensing designs to other companies, PAX Water's business model is to build and sell its own water mixing equipment.

PAX Lily mixing impeller

Its patented mixer, literally modeled on frozen whirlpools, is now being used by more than 200 cities to mix and destratify layers of stagnant water in over 500 large-scale drinking water storage tanks. As a result, water coming from your tap can have as much as 80 percent less disinfectant residue (including the carcinogenic chlorine you sometimes smell and taste) and the municipality can reduce the energy it takes to keep its water fresh by up to 90 percent. PAX Water has doubled its sales every year and has won a number of industry awards. It's recognized as a leader in its scientific credibility and is now introducing other biomimetic products to the potable water market.

Following the launch of PAX Water, we decided on a hybrid model for our next subsidiary, PAX Mixer. PAX Water had chosen to target a well-defined emerging market with great growth potential. PAX Mixer's charter was much broader—all mixing (except water), including food, chemicals, paints, petroleum, medicines, blood, and so on. PAX Mixer's founding CEO had experience with accessing government funding; we were soon awarded a three-year multimillion-dollar grant from the Department of Commerce's Advanced Technology Program to study the benefits of mixing with our biomimetic impellers and in-line mixers. This was an extremely productive and positive experience and the first of many grants for PAX companies.

By early 2007, PAX was cooking. Contrary to a handful of years before, oil prices had now soared, clean tech was the new dot-com, and venture capitalists were eager to meet with us. We were continually assessing how to best grow the business. With three sectors launched and hundreds to go, we realized that we would all be long dead if we kept starting companies one at a time. We felt we had enough applications ready to launch a bigger effort, where several, high-energy-using markets could be tackled at once. For example, the market sector called HVAC (heating, ventilation, and air-conditioning) uses more than one third of all the electrical energy generated in the United States. Unlike a fan blade or mixer, an air-conditioning unit is a system made up of components and subsystems like fans, pumps, ducting, and heat exchangers. Developing entirely new systems meant developing several PAX technology components in concert. That meant big money, and big money meant some form of venture capital.

After months of discussions with our advisers and shareholders, we partnered with an entrepreneur and venture capitalist, billionaire Vinod Khosla. After many years with one of the leading venture capital firms in the United States, Vinod established a new venture fund with a focus on clean tech. Khosla Ventures was investing hundreds of millions of dollars in technologies ranging from ethanol fuels, to energy efficient lighting, to more effective battery storage. We raised almost $13 million, including funds from our own shareholders who wanted to expand their participation, and spun out our next subsidiary, PAX Streamline, in early 2008. PAX Streamline set up a research facility, hired a top team of researchers and business staff, and began to expand into six product areas, including wind turbines, pumps, and a novel refrigeration cycle. Outside the walls, storm clouds gathered in United States and worldwide financial markets.

The inventions of refrigeration and air-conditioning, around a hundred years ago, have profoundly changed the world. For millennia, humans preserved their food by fermentation, salting, drying, smoking, or pickling. In fact, it's thought that beer was originally invented to preserve grains that otherwise would have become moldy or been eaten by rodents and weevils. Entire armies, needing to take preserved food on the march, were only as good as their supply of the salt that preserved their protein sources of meat. Many military campaigns, including those of the Revolutionary War and the Civil War, implemented strategies to attack and seize, or at least disrupt, the enemy's salt mining operations. No salt—no food—no march.

Refrigeration invented by Willis Carrier changed all that. However, with it came serious side effects. Refrigerant gas, such as ammonia, is highly poisonous. Other commonly used gases have severe, detrimental impact on the atmosphere. Some have a global warming potential (GWP) thousands of times higher than carbon dioxide. On top of these problems, the conventional refrigeration cycle is a major energy user and therefore a large contributor to carbon dioxide emissions from the associated power generation. It would be of great benefit to human society and the environment if we could solve these problems.

Nature is an expert at refrigeration. Animals perspire to cool themselves. Our personal supercomputer, the human brain, only gets hot when it is ill and only by a few degrees. Tornadoes and hurricanes are essentially heat engines that refrigerate with high efficiency and no deleterious gases. Decades before, I'd identified nature's operating principle for cooling and developed a concept for a new approach to refrigeration and air-conditioning. Our research team at PAX Streamline, led by Dr. Tom Gielda and assisted by Dr. Steve Eckels and his team at Kansas State University, applied rocket-science physics to the concept—and our prototypes worked. It's been called the first new refrigeration cycle in one hundred years.

Our products were gaining traction as climate change and global pollution were getting more and more press. Traveling to Europe and Asia to speak at conferences, I learned of increasing problems in megacities caused by smog. I became strongly interested in applying our mixing technology to a very large market—the sky.

Pollution is usually carried away from cities by wind, but a number of huge cities with the worst air quality have grown up in regions with strikingly similar topography—ringed on several sides by mountains. This can create a bowl full of choking smog that, depending on weather patterns, just doesn't blow away. Los Angeles, Beijing, Mexico City, Milan, Tehran, and dozens of other cities experience pervasive, health-destroying pollution for much of the year. Could we use the proven effect of PAX geometries to safely eliminate dirty inversion layers from cities and dissipate smog, without negative side effects? Our computer simulations said yes. As well as nudging the smog to move up and out of the bowl surrounding it, it appeared that this biomimetic approach also had the potential to nucleate raindrops in humid areas. Large, arid regions of the world often have very high humidity—up to 90 percent—but no rain. Places like the Gulf States in the Middle East, northwest Australia, and parts of Africa and China all have high rates of evaporation from warm, nearby seas. Excited by the prospect of making a large-scale difference, we contracted an entrepreneur who was game for the adventure and asked him to explore the potential of taking this idea to reality. He spent several months meeting with climatologists, politicians, and businesspeople in the United States and China to assess the pros and cons of our concept. Through his network, he met a UK businessman who had started successful clean tech and carbon trading companies in China. He currently lives in and loves Beijing, but the smog is often so severe that he questions whether his family can continue living there. In a move that showed true entrepreneurial spirit, he sponsored PAX Mixer's research on atmospheric mixing. Experimentation went well, and by the end of 2012, we had completed compelling initial research. The next step will be to obtain governmental support for field testing, perhaps in China. We'll all have to stay tuned for the rest of the story.

As a boy, I had the privilege of realizing that nature *only* moves and grows in precise, turbulent, spiraling flows. As an adult, I learned that human technology, in the main, tries to suppress turbulence. Nature doesn't waste the opportunity. It exploits the energy that is rolled up in turbulence. Birds, insects, fish, and the human heart clearly demonstrate the advantage of this strategy. Humans insist on traveling in straight lines and guzzle energy. Nature travels in spirals and sips energy. Truly grasping the significance of this simple fact throws open the door to reinventing the industrial world and gives us the tools to rescue our ailing planet, populations, and economy. By adapting and applying nature's spiraling geometries, I am confident that we can halve the world's energy consumption—without sacrifice.

Chapter 3

CATCHING THE WORLDWIDE WAVE

B y building products and businesses that can maintain themselves without degrading any resources or generating waste, sustainability is biomimetic at its core. In both government and industry, the economic benefits of sustainable products and alternative energy are proving their worth. Clean tech is less than ten years old as a recognized market sector, but it is the fastest growing venture capital investment market. During the midst of the Great Recession in 2010, venture capital investment in clean tech was still $7.8 billion and in biotech was $5.4 billion. While currently seen to be a subset of both these sectors, research suggests that venture capital investment into biomimicry as its own category could eclipse clean tech and biotech in the years to come. Some argue that this new industry could outpace other huge sectors to rapidly become a $100 billion market. I know of biomimetic technologies just in the fields of refrigeration, construction, and water desalination that, alone, hold that potential.

From improved urban planning to increased crop production to renewable energy generation, cities and countries around the world are also taking real steps to protect and live in harmony with nature. I spoke alongside Kenny Ausubel, the cofounder of Bioneers, at forums in the Netherlands and Hong Kong in the past couple of years and was deeply impressed by the level of climate action, clean tech development, and sustainability activity in Europe and China. Bioneers was started in 1990 as an annual conference where pathfinding scientists and social innovators could share breakthrough solutions to major environmental and social challenges. It has grown to become a multifaceted, international organization offering influential conferences, educational networks, and cutting-edge programs in media-based public education, food security, resilient communities, indigenous rights, and leadership development, all "based in a philosophy that recognizes the aliveness, interdependence, and intelligence of the natural world." Kenny and his wife and cofounder, Nina Simons, are tireless advocates for people working around the world to implement critically needed solutions.

As Kenny said when we spoke recently, "We're entering an ongoing state of large-scale breakdowns. The trick is to turn breakdown into breakthrough by reimagining civilization in the age of nature and making that transformation happen very rapidly. The years running through about 2017 are the once-in-a-civilization window for a massive global shift. In response, the biomimicry and design science revolution to restore people and planet is well under way and it's absolutely fundamental. The solutions are largely present, or at least we know what directions to head in. At this point, however, the United States is losing the clean-tech high ground to Europe and China. They're developing and implementing breakthrough clean technologies at a rapid pace—for instance, the EU has set a target to capture 25 percent of the global green products market, while China is innovating at a state level on biomimicry, clean tech, and large-scale ecosystem restoration. If the United States fails to get on track soon, we'll never catch up. At the national level, it's going to come down to an issue of business competitiveness, as the former Defense Secretary Robert Gates and Energy Secretary Steven Chu have warned."

The European Union is certainly ahead of the United States in legislating for sustainability. Germany, the Netherlands, the Scandinavian countries, and Switzerland frequently top the list of green countries. The people of Denmark, for example, recycle or incinerate 95 percent of their waste, while 54 percent of U.S. waste goes to landfill.

One rapidly growing international organization is implementing sustainability as a response to climate change—and already represents 8 percent of the world's population. The fifty-eight megacity members of the C40 Climate Leadership Group span a huge range of languages, wealth, cultures, and infrastructure, yet despite their differences, these communities face many similar challenges. C40 was founded in 2005 by a group of mayors to work collaboratively across geopolitical lines—and while international negotiations on climate change continue to make erratic progress, the C40 cities are galloping ahead. As of mid-2011, they had collectively already taken nearly five thousand actions to reduce their carbon footprints and address sustainability issues, with another fifteen hundred under way.

Specific to biomimicry, there is already one city that has made a public commitment to becoming a hub for its development. San Diego, California, has established the Biomimicry BRIDGE (business, research, innovation, design, governance, and education) as a partnership between local universities, business development groups, and San Diego Zoo Global. As part of this effort, the zoo has also launched a Centre for Bio-inspiration to support technology incubation and transfer. The city recognizes that public-private partnerships will be the most effective way to get biomimicry out of research labs and into profitable products that will sustain their local economies.

There's an old marketing saying that humans are driven by fear or greed. While crude, it makes sense. We're hardwired to avoid what seems dangerous and crave what looks tasty. Businesses use this to sell their products, and so do governments, which use tax breaks and other incentives to pull citizens, businesses, and municipalities forward into new habits, and regulations with penalties to push them away from bad ones. It can be an effective combination, but it's only part of the puzzle.

GREEN VERSUS INDUSTRY

Beyond valuing different aspects of the bottom line, environmentalists—sometimes slighted as impractical tree huggers—and big business have increasingly been cast as enemies. Such a binary view blinds us to the potential for collaboration. His Royal Highness the Prince of Wales emphasized this dynamic and examples of success in his 2012 film, "Harmony: A New Way of Looking at Our World." Since the first industrial revolution, businesses worldwide have developed highly effective processes for everything from manufacturing to distribution. Biomimicry does not propose to overturn existing ways of doing business but rather to build on and optimize them in order to reduce waste and negative side effects. That in turn increases profits. Now many of the world's largest companies are jumping on the nature-as-mentor bandwagon. Examples include Honda's "Environmentology"; BP's "Beyond Petroleum"; and GE's "Ecomagination." Are corporations sincere, posturing, or both? Unfortunately, the reality is often more sizzle than steak.

"Greenwashing" differs from truly sustainable corporate practice. The term, coined by environmentalist Jay Westerveld in 1986, refers to reframing, repackaging, or relabeling products or processes, to make them seem environmentally friendly to consumers. The reasons for greenwashing are twofold: to increase profits and to weaken the progress of restrictive regulations. With three-quarters of the surveyed public stating that they prefer to purchase products from clean and green companies, it's obvious why there is such an imperative for businesses to be seen this way. Unfortunately, there are few, if any, standards that greenwashers are held to.

In 2009, TerraChoice, an independent, nonprofit Canadian firm affiliated with Underwriters Laboratories, investigated 2,219 common consumer products and found that over 98 percent were greenwashed. Examples ranged from "bamboo" fabric (most is actually chemically treated rayon) to "certified organic" shampoo with no organic certification to "energy-efficient" electronics containing hazardous materials to "environmentally friendly" pesticides to "natural" cotton that was full of pesticides and formaldehyde.

The challenge of greenwashing is that it makes it difficult for people to differentiate between sustainability and lies. The media is so filled with reassuring claims of clean coal, safe "fracking," clean cars, and natural foods that the general public can allow itself to believe that all is good—the problems are being taken care of. Fortunately there is a growing, although slow, recognition of the need to hold product and service providers to a standard of truth in green advertising. In this age of the Internet and social networks, public awareness can grow rapidly.

Any dinosaur company misrepresenting itself will eventually experience negative consequences as a result of the public's increasing discernment.

Governments are acknowledging their role in ensuring accuracy in sustainability. The Australian government has now legislated fines of up to $1.1 million for misrepresentation in environmental claims. The U.S. Federal Trade Commission is retooling its green guidelines and, little by little, starting to create a tougher environment for false claims, though there is continuing pressure from corporations to deregulate. The Canadian Standards Association discourages "vague claims" and insists on proper data. Norway has banned its auto industry from claiming that cars are green, clean, or environmentally friendly. As Norwegian consumer ombudsman, Bente Øverli, stated bluntly, "Cars cannot do anything good for the environment: except less damage than others."

How can you determine if a product or company is greenwashing? As a popular 1970's bumper sticker read, Question Authority. Remember that product marketing exists to promote, not to educate. Get the facts behind the facts. Use the Internet—for many of us now only as far away as our cell phones—to understand what terms mean.

It's too much work to become an expert on everything you buy, so there are organizations that are doing the research for you. Whether food brands, personal care and cleaning products, or even politicians, there are growing numbers of websites that will give you report cards including Underwriters Laboratories' (TerraChoice) SinsofGreenwashing.org, Consumer Reports' GreenerChoices.org, and Greenpeace's Stop GreenWash.org. Of course, such resources can also be greenwashed. One website that I recently found shared information on chemistry as a teaching aid for schools that "scientifically" asserted the complete safety of chemical manufacturing. It seemed quite independent and academic. Only by running the lead author's name as a separate search and following a few leads was she revealed as a senior scientist for one of the world's largest, nongreen chemical manufacturers.

Nutritional labels, an oddity not long ago, are now standard. Energy Star labeling, which describes efficiency in U.S. appliances, has been well received, and there are corollary labels in many other countries. Well-respected organizations are also responding to the call for labels and certifications that quickly signal to a consumer whether a product is truly green. In early 2012, the International Organization for Standardization proposed a technical committee to create biomimicry standards for European markets. Biomimicry 3.8, the world leader in biomimicry education, is considering a mark that confirms a product is designed and built in accordance with bio-inspired principles. PAX is investigating the establishment of something similar for engineered products. The good news is that truly biomimetic products don't need to greenwash, because they are 100 percent green by nature.

NATURE'S TEACHERS

Biomimetically inspired products will be found in almost every corner of the marketplace, from medicine to aerospace to manufacturing to transportation within the next ten to twenty years. The impact of these innovative goods will further change the way we think about our human environment—not only in how we design things like homes or offices but also in how we conceive of entire communities as ecosystems, including businesses, government bodies, and other social organizations. Let's look at an organization that's leading the way.

Janine Benyus, author and biologist, is a true pioneer with a rare gift of humbly presenting nature's technological wonders. She and her business partner, Dayna Baumeister, established the Biomimicry Guild in 1998 with a goal of bringing biologists to the design table. With more than 250 clients since then, including many Fortune 500 companies, the guild consults to multinational companies, governments, and universities. They've helped redesign furniture, sneakers, carpets, manufacturing processes, airplanes, and even entire cities. In 2006, Janine and her colleagues opened a sister firm, the Biomimicry Institute. The institute reaches millions through talks and has trained hundreds of biologists, designers, and businesspeople who now practice biomimicry in their professions. The training offered by the institute has resulted in a network of K–12 teachers and university professors who are teaching biomimicry to the next generation. As part of this educational outreach, in 2008 the institute launched AskNature.org, the world's first online library of nature's solutions. It's a remarkable resource, growing daily, that cross-references thousands of bio-inspired solutions. Information on Ask Nature is set up under a classification system—the biomimicry taxonomy—organized by how organisms meet different challenges. Strategies used by champion adapter animals and plants are posted by research scientists worldwide. The database is organized by how organisms "break down; get, store, or distribute resources; maintain community; maintain physical integrity; make or modify their own forms; move or stay put; and process information." Users can update, refine, and connect with one another to further develop bio-inspired design.

Over the years, the Biomimicry Institute, guild, and its growing outreach projects have become an ecosystem in themselves. To streamline operations, activities were combined into one umbrella organization, called Biomimicry 3.8, which encompasses the nonprofit institute, a corporate consulting firm, a worldwide speakers bureau, and a highly regarded professional training and certification system.

Janine is the first to say that there is an extraordinarily dedicated and passionate team behind the scenes of her work. Those who have met Dr. Dayna Baumeister, cofounder of Biomimicry 3.8, know her to be one of the world's most qualified, knowledgeable, and articulate advocates for biomimicry.

As Dayna told me in a recent conversation, "I see what we humans as a species have done to this planet and it stabs me in the heart." In response, she is passionately committed to enabling and fostering the practice of biomimicry in early adopters and innovators around the world, so it becomes a standard practice in all human endeavors. From their years studying the natural world and their work with companies in diverse fields that are striving for more sustainable models, Dayna, Janine, and their colleagues have a deep understanding about the challenges we face as a result of our disconnect from the natural world—a disconnect that allows us to damage the very environment that supports us. Why has this happened? Dayna believes that humans have become less aware of subtle connections as we have moved to more and more urban, structured environments. Until very recently, the vast majority of us ate food that we grew or caught ourselves. We were intimately aware of seasons, animal behavior, plant growth, and how they intersected. If you remove humans from immersion in such a complex, interdependent environment, you soon have a generation of children who sincerely believe that milk comes from a factory and cereal from a box. The other challenging problem with the human psyche, in Dayna's view, is our resistance to change. We can cling blindly to a sinking ship, even if there are clearly ways to save ourselves.

I asked Dayna if she could name three or four large corporations that have made a real commitment to biomimicry. There was silence for several seconds. "Large U.S. companies?" she asked, and paused again. "No, not really. In this recession, most companies know that they should be investing in future technologies, but they're reverting to the tried and trusted, albeit inefficient ways of the past." Dayna did highlight the Brazilian beauty and health care company Natura (something like Avon in Latin America). Natura has made a long-term commitment to training its staff in biomimicry and applying this to its products and services. She also praised the long-standing efforts of Interface, the world leader in carpet tiles that we'll meet later in the book. Qualcomm, a leader in wireless technology, has also invested strongly in biomimicry by developing an electronics screen based on the refractive abilities of butterfly wings. Dayna believes that there are champions and a foundation of interest in a number of large companies, and she expects that biomimicry will increasingly gain ground in larger institutions as the worldwide economy improves.

Dayna and her team at Biomimicry 3.8 see much more effort and commitment from smaller, flexible companies that don't have a legacy of growth built on old models. As Dayna explained, a number of Fortune 100 companies have started down a path toward bringing biologists to the design table and biomimetic practices to their business. Often there is an internal champion who pushes hard; but, even given that support, it sometimes doesn't percolate through to real institutional change. Wall Street's quarterly performance goals and the inertia created by multi-layered organizations result in slow or even abandoned efforts. Dayna recently met with one of the top Ecomagination staff from GE: "But they weren't looking for ideas from outside their own walls, let alone from nature."

Procter & Gamble is a leading household products company that does have a strong culture of seeking outside ideas, including biomimetic solutions. In part because biomimicry doesn't yet have long-term metrics in Procter & Gamble's field that prove its return on investment; solutions haven't yet left internal research and development to become retail products. Many companies won't adopt a technology until it has been in the marketplace for several years—a real chicken-and-egg situation.

While retail products are ramping up, Dayna is happy to see the increasing uptake of biomimicry into architecture, because the built environment is particularly fertile ground for its benefits. As she explained, "Since buildings account for about 50 percent of total U.S. energy use, the greatest collective impact could come from applying biomimicry to buildings, communities, and cities."

The Biomimicry Guild has an alliance with HOK, one of the world's largest architectural firms. When the firm takes on a new project, a team of HOK and biomimicry professionals visits the site to research its "champion adaptors," in order to customize their architectural approach to the most successful strategies used by plants and animals of that region. Collaborations with HOK include a South Korean skyscraper that requires fewer building materials by mimicking the stable, lightweight structure of a honeycomb. They're also applying biomimicry to Langfang, a city near Beijing, China. Four thousand years ago, Langfang was a forest, which captured and distributed rainfall. Currently, the built environment treats all that rainwater as runoff to drains, while the inhabitants draw their daily water from declining aquifers. HOK has redesigned the city's architectural plan with strategically planted belts of trees in the ancient watercourses, to once again replenish aquifers.

Dayna emphasized that the main challenge for companies deciding whether to adopt biomimetic solutions hinges on value generation. Profit is usually the only metric that is used, and while she recognizes the tremendous potential for profit offered by biomimicry, she stressed that there are also highly valuable, albeit less easily measured, benefits for companies that adopt biomimicry into their practices. Employees see real purpose and personal mission in their work. It creates passion, loyalty, creativity, and team building. Biomimetic product development starts from a nontoxic, nonharmful stance. Rather than designing for end effect and then compensating for toxicity and waste management, it also saves adopters considerable money on increasingly arduous and expensive environmental regulations—and future remediation liability.

HUMAN NATURE

As we'll see throughout this book, nature's problem-solving strategies offer immense opportunities for wealth creation in a new economy. Whether the shatter-resistant ceramics of an abalone shell, the superior drag resistance demonstrated by seaweeds, or the pure combustion of plant photosynthesis, impacted industries range from construction to transportation to medicine to software. Nature's paradigm does what human technology tries to do—and does it sustainably without stripping out the base resources that create wealth. So what's preventing us from more rapid adoption of bio-inspired design?

As Dayna Baumeister suggests, if we're going to succeed at stabilizing and growing world economies, while reversing the decline of the planet, we must be aware of our human nature. Nature has given us an irrepressible urge to protect and improve our physical circumstances. This underlies our phenomenal progress as a species but also the human-created problems we're facing today, because that powerful instinct for personal and familial survival can distort to terminal greed, where we lose touch with reality and see everything as "If you win, I lose."

Of all the species on earth, we seem to be the only ones lacking an "enough" gene. In the wild, dogs, lions, cows, monkeys, apes, even mosquitoes and houseflies, eat until they are satisfied. They don't keep eating to obesity. Animals from squirrels to blue jays store food for winter—and some do store a little more than their needs. This might be seen as suboptimized evolution, as if they weren't evolved enough to remember where they'd hidden all their stores. However, the leftovers benefit other animals and move seeds to new growing sites. It's all part of a balanced ecosystem with zero waste.

The attitude of "never enough" can permeate every human endeavor.

I knew a man in London who sold his business for one hundred million pounds. It had long been his goal and dream to have exactly that amount. I didn't see him again until several years later, when we had lunch together. He confided in me that he was being treated for clinical depression and had been living in hell for the past two years. I asked what had happened to leave him in that state. He told me that he'd been over the moon with his hundred million pounds and had had a marvelous time buying a villa in the South of France, a Ferrari, and a yacht. Then one day he realized that he had spent eighteen million pounds and no longer had the hundred million of his dreams. He panicked and put all the rest into blue-chip bonds and savings accounts. The interest he was earning was not keeping up with his cost of living, inflation, and taxes, so his pile was diminishing. Here was a man who wouldn't spend all this money in a lifetime of luxury, but was tortured.

Unbridled desire—the more, more, more syndrome—shows up over and over in humans interacting with the material world. We all want more—more money, more security, more luxuries, a better car, a bigger house, new clothes, to be more famous, a better job, a better spouse, a better nose or boobs, to be more enlightened, or for the world to have more peace—or more sex, more power, more hair, better weather. Everyone wants something he or she doesn't have and goes to great lengths to get it. And when we've got it? We want something else. We alone, among all the millions of life-forms on earth, seem to spend virtually no time satisfied. This appears to be a defining characteristic of modern humanity—the incessant, seemingly obsessive, often illogical striving for more.

There are exceptions. It's widely described that Australian Aboriginals didn't accumulate personal possessions: Anything and everything belonged to the community. Aboriginals appear to have lived this way since prehistoric times fifty thousand years ago. Even when confronted with Chinese traders who sailed down to Australia from Asia over the past thousand years, Aboriginals chose not to copy any of their technologies, artifacts, or societal norms such as ownership. Instead, they stuck to their ancient roots.

A friend who lives in Micronesia told me that the inhabitants of one of the islands of Polynesia were given refrigerators by a group of visiting, well-intentioned missionaries. They had noticed that the locals, who were subsistence fishermen, had to fish every day because any excess catch spoiled in the tropical heat. The missionaries thought it would be a blessing if excess fish could be refrigerated, allowing the fishermen to put their attention to other wealth-generating activities. On a return visit a year later, the missionaries noticed that there was no trace of the refrigerators in the community. Their inquiries informed them that the elders had ordered all the equipment dumped in the ocean. The reason? Refrigerating excess fish meant that surplus was no longer given to the elderly or infirm, as had been their custom for a thousand years. It was unacceptable to the tribe that "progress" resulted in more wealth for some and hunger for others.

If indigenous peoples have a sense of "enough," what has happened with the rest of humanity? Many of us in the developed world have tamed and caged and bored ourselves. Like animals domesticated for use, we have become fat and unhealthy. With more than half of us living in urban areas, we've largely lost our connection to nature and the historic initiation rites that oriented us to our place in the cycle of life. Instead, we distract ourselves with everything from shopping to stimulants to video games.

A new green economy can easily suffer from the same predatory form of capitalism that created the recent global economic meltdown. As Kenny Ausubel of Bioneers notes, "The world is suffering from the perverse incentives of 'unnatural capitalism.' When people say 'free market,' I ask if *free* is a verb. We don't have a free market but a highly managed and often monopolized market. We used to have somewhat effective antitrust laws in the United States. Now we have banks and companies that are 'too big to fail,' but in truth are too big not to fail. The resulting extremes of concentration of wealth and political power are very bad for business and the economy (not to mention the environment, human rights, and democracy). One result is that small companies can't advance too far against the big players with their legions of lawyers and Capitol Hill lobbyists, when in truth it's small and medium-sized companies that provide the majority of jobs as well as innovation."

Capitalism exists, in large part, because of the lack of an "enough" gene. In fact, a mainstay of the so-called American dream is that our children will have more than we did—no matter how much we have—but we can't keep having more and more in the ways we're used to without destroying ourselves. After all, what possible use does nature have for a billionaire? Which of the millions of species on earth, apart from humans, has the equivalent of even a millionaire? I've been told that if we were to add up the wealth of the top 1 percent in the world today, we would find that it is greater than all the wealth owned by all the earth's peoples combined before the twentieth century. According to Forbes, the world's twelve hundred and twenty-six billionaires—a group that totals far less than 1 percent of the world's population—have assets exceeding the combined annual incomes of nearly half the world's population—a net worth of $4.6 trillion. The top 10 percent of humanity owns 85 percent of the world's wealth today. This statistic probably hasn't varied much throughout the history of civilization.

"I didn't know you dealt in dinnerware," I said quizzically to my host, Wolfram. I was visiting my friend, an upmarket jewelry dealer in Zurich. He had taken me into his large back room to have coffee while we talked over old times. There were several carved-leg dining tables covered with fabulous collections of antique fine china and silver and gold flatware. The settings were laid out over exquisite brocade tablecloths.

"I don't, normally," he replied, "but some of my best customers for jewelry and watches are Middle Eastern oil billionaires. They asked me to find these items for them—and it's quite profitable. There's over a million dollars' worth here."

"Wow," I exclaimed. I could imagine that many of these works of art had graced the tables of European royalty and nobility.

"You'll never guess what they do with them," Wolfram continued. "They take them on their superyachts and use them for picnic settings with people they want to impress. When the meal is finished, they bundle up all the settings in the tablecloths and throw the whole lot overboard. It's their equivalent of disposable paper plates."

Let me be absolutely clear. I have nothing against capitalism and people enriching themselves and their families, but a rapidly growing number of business leaders, politicians, and scientists agree: The global climate, both physical and economic, is changing and it will greatly impact our lives. Change is essential and the benefits should be convincing, but we're hampered by our hardwired, often distorted, hunger for survival and resistance to anything new. How can we use the knowledge of our resistant nature so that it serves us, the collective us, and our survival? In order to survive, we need to use our highly evolved minds to collaborate with our instincts. We can teach our hunger for abundance that sustainability does not have to mean sacrifice. For instance, driving a more energy-efficient car costs no more and saves both fuel resources and our atmosphere, without pain. Replacing highly inefficient incandescent lightbulbs with state-of-the-art LEDs or CFLs may cost a bit more at the outset but saves multiple times that money and energy over the bulbs' years of use.

Kenny Ausubel has observed the response to Bioneers' environmental and social innovators for more than twenty years. As he says, "It sounds cynical, but many believe—and there's lots of historical precedent—that people simply will not change until forced to. Climate change is giving us an environmental education—the hard way, and the perfect storm of environmental and social crises that's unfolding is producing a civilizational endgame where we'll have no choice but to change. We also know that people are capable of amazing acts under dire circumstances, and we can come together astonishingly well to cooperate to meet these kinds of cataclysms. Already people are rising up all around the world to make systemic change. They're reacting to the extreme concentration of wealth and power, to environmental collapse, and the lack of social justice—the same as we saw in vast popular and political movements in response to the age of the robber barons more than a hundred years ago. In the era before the invention of public relations, the infamous railroad robber baron and monopolist Collis Huntington candidly stated his business credo, 'Anything that is not nailed down is mine; and anything that can be pried loose is not nailed down.' It sounds like a basic corporate mission statement."

Kenny emphasizes, however, that "Worldwide the revolution has begun; from the Arab Spring and Occupy Wall Street to the Green movement in Iran, the freedom movement in Syria, populist risings in Myanmar, and now even in Russia. The critical mass is rising, and with social media's ability to organize and communicate, the tipping point is inevitable. It's not a matter of if but of how soon the 'shift hits the fan,' as Tom Shadyac put it. There is unquestionably a global convergence toward a new paradigm. Our challenge is to accelerate it into fast-forward and shine light on the practical solutions and strategies that we already know can succeed."

The present driving paradigm of man is the philosophy of science. Science is valid. It works. We have cars and computers and heart surgery. We can make new molecules and talk via satellites and travel around the world in a day. Yet it wasn't long ago that many of us would have used wood or coal for heating and cooking, pickling, drying, or salting to preserve food, horses or donkeys for transport, and candles or whale oil for lighting. Given the rapid change in the basics of our own lives, we must intellectually realize that there will always be a new paradigm, and another and another.

The scientific truth in Europe only a few hundred years ago was that the earth was flat and had edges to fall off. This was well considered and fully proven, based on the evidence available. In fact, you might have been burned at the stake if you argued against it. Under this paradigm, all Europe was explored and mapped and its geography was understood. Totally valid work was conducted and, in its context, was not wrong. The questions of the time were confidently and largely answered even if erroneously. However, that dominant paradigm precluded the discovery of greater knowledge. Once challenged, the results provided vastly more opportunities for Europeans. For example, it allowed Christopher Columbus to "find" the Americas. (Of course, native peoples had been well aware of the New World for millennia and didn't regard it as "lost" in the first place.)

The prerequisites for further advancement of science and technology leave us in something of a quandary, since we must first accept that "laws" may not be infallible after all. Recently, researchers at the Gran Sasso National Laboratory in Italy reported measuring neutrinos traveling faster than the speed of light. Even if not confirmed, it shows that some leading scientists are willing to challenge our most basic assumptions. If we're truly objective, we have to be willing to look for new sets of rules, while not knowing entirely what it is that we're looking for. All this must be done while remembering that challenges to the status quo are very unsettling to many classically trained philosophers, physicists, and engineers. There usually follows a protracted squabble as learned intellectuals insist that the new ideas are not to be taken seriously.

When President Rutherford Hayes heard about the invention of the telephone, his response was,"That's an amazing invention, but who would ever want to use one of them?" The president of the Western Union Telegraph Company called the telephone "an electrical toy" and declined the opportunity to buy the patent. It was publicly declared by the world's foremost engineers that it was impossible for machines heavier than air to fly—just before the Wright brothers proved them wrong. In the 1970s we saw the powerful Swiss watch companies dramatically lose market share after watchmakers resisted the quartz watch "gimmick." Kodak similarly dominated its industry—until it chose not to invest in emerging digital camera technology.

History has many examples of such intellectual buffoonery. It's easy to dismiss our ancestors as small-minded, superstitious, or at least misguided in their theories, yet future generations will look at us the same way not because our theories don't serve a purpose, but because in every age people believe the knowledge of the time to be much more true than previous understandings. When we can say, in all humility, that we still don't know everything, and that there may be better ways of achieving our goals, we can look with fresh eyes and discover new and more relevant paradigms.

Biomimicry offers the opportunity to meet our resource needs and to reinvent almost every industry on earth. But as we all learned in school, an object at rest wants to stay at rest, and an object in motion doesn't like to stop or change direction. Like all engineers who must deal with those two fundamental laws of physics (as yet still proven), innovators must cope with this corollary to our survival instinct: resistance to change and the resulting inertia created by systems and institutions that are already in place. The trick is to find the path of least resistance. All it takes is for each of us to be willing to recognize our human nature and take ourselves in hand. We are voting every day by our action or our inaction, by what we buy and what we talk about. Whether by supporting biomimicry education in our schools, speaking up for a biomimetic project or practice in our businesses, showing up for a city council meeting on sustainability, or researching the products we buy, each of us can be a tremendously powerful force for positive change.

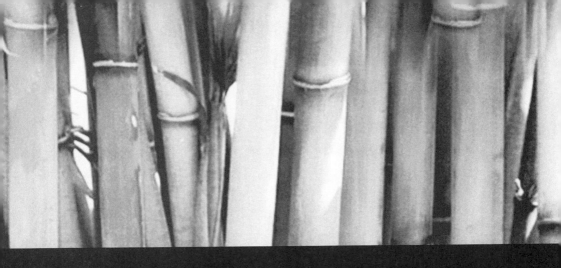

Part Two

BIOMIMICRY AT WORK

Chapter 4
....

SECRETS FROM THE SEA

We were anchored in the remote Sholl Island passage off Australia's northwest coast. It was 1970 and I was a twenty-year-old Fisheries and Wildlife officer, spending weeks at a time patrolling the Indian Ocean. On board were five wildlife research personnel that my skipper and I were hosting on the first-ever wildlife survey of the Dampier Archipelago. These rugged islands, largely made of huge iron ore boulders, towered to several hundred feet. Sparse vegetation grew between the boulders. Spiky spinifex grass; tangled, stunted trees; and the shear-sided rocks made these islands almost impossible to explore. A large tidal range, strong rip currents, brown ocean made muddy by miles of mangrove swamp, venomous sea snakes, biting sand flies, and blistering hot sun all combined to ensure that this area of the world had not been spoiled by tourism.

It was about nine o'clock on a moonless night. We had eaten dinner and I, as was my custom, stood on the back deck, tending my heavy-gauge fishing line with which I would often catch big fish. With a twang, the line was suddenly ripped out of my hand by a locomotive-size force. Fortunately, it was tied to the boat's guardrails. I yelled, "Bring the searchlight" to the men belowdecks; and a few seconds later, light flooded a large area behind the boat. "Holy crap!" someone yelled. "Look at this!" The ocean behind us was boiling with sharks. There must have been twenty or thirty, each measuring around eight to ten feet long. A shark had taken my bait and hook and now the other sharks were attacking it, ripping dinner-plate-size chunks out of its body.

Gary, a wildlife researcher unused to boats, came bounding up from belowdecks to see what all the excitement was about. As he hurried, he tripped and his momentum took him through the guardrails, over the side of the vessel, and into the water. On his very brief journey, he no doubt glimpsed the welcoming feeding frenzy he was plunging into.

To the amazement of all of us watching, a dripping-wet Gary was suddenly back, standing on the deck, just as quickly as he had gone into the water. Somehow, the sight of all those thrashing sharks had pumped him to superhuman strength and speed. He had grabbed the edge of the deck with one hand and somehow managed to pull himself out of the water. Luckiest man I ever saw. From then on, I never witnessed Gary on deck at night.

DESIGNED FOR SPEED

Sharks have been around, biting things, for almost a half a billion years. There are 440 species that range from six inches long to the biggest fish of all—the forty-foot whale shark (which actually has no relation to a whale). It's a little ironic that while they're notorious for it, sharks very rarely eat humans—or even bite them. Only about seventy-five shark bites are reported worldwide every year, with an average of four deaths. But humans eat sharks—up to one hundred *million* of them per year. In fact, just their fins can sell for more than $300 per pound to make shark fin soup. It's one of the world's highest priced ingredients.

What do sharks have to do with business innovation? Recently, plenty. They're being used as models for a number of new, extremely valuable products impacting everything from hospital safety to ship hulls to Olympic medals, thanks to the shark's evolutionary need to keep moving. Unlike other fish, most species of sharks can't activate their gills to extract oxygen from the water. As a result, they're compelled to be constantly on the move to keep water flowing through their gills—even while sleeping. All this movement costs a lot of energy, which then has to be supplied by the rewards of successful and vigorous hunting. Nature always likes to minimize energy use by living organisms; the most efficient animal runs the fastest and farthest on the least amount of energy and therefore survives better than less-efficient members of its family. That's valuable evolution. Sharks are no exception and, in fact, are prime examples of extraordinary, streamlined design—more so than any human-designed object in a number of ways. How do they do it? A shark is somewhat bullet shaped, which reduces its profile as it moves through the water. Its fins and tail muscles work in concert to create water vortices in their wake, which the shark can, in effect, lean against as it propels itself forward. Finally, it gains drag resistance thanks to its cleverly evolved skin, which is made of tiny vertical scales known as placoid scales, or dermal denticles—little ridges with quite a rough feel to them. They are so rough that before the advent of modern sandpaper, carpenters used sharkskin for sanding wood.

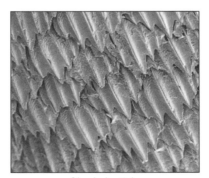

Shark denticles

It seems counterintuitive that a rough surface offers less drag, or resistance, than a smooth one. However, nature proves this to be true in countless examples. Imagine dragging your arm through the water in a bathtub or swimming pool. You can feel the pressure on your arm as the water molecules flow around the obstruction (your arm). What is harder to feel, but is also occurring, is that your arm is dragging water along as well. In a similar way, a ship can drag its own weight in water as it travels. Pulling the ocean along with it obviously means that the ship burns much more fuel than if it could leave all that water behind;

added to which, the boat's propeller operates, in part, in that same water that's being dragged. This reduces propulsion efficiency. Ships' operators worldwide traditionally try to keep their hulls as clean and smooth as possible to reduce drag. A shark's skin is rough; but rather than making it less slippery, its surface actually keeps large amounts of water from sticking to the animal and dragging against forward motion.

Essentially, by roughing up the water right at the intersection of the shark and its surroundings, less of the water sticks to the shark to slow it down. This feature is currently being exploited by a number of businesses. It's an example of biomimicry based on form, where a particular mechanical shape, evolved by nature, results in improved performance when applied to engineered devices or tools.

Case in point: Inspired by the shark's dermal denticles, researchers in Germany have developed a paint that, when applied with a particular stencil, creates ridges in a pattern that improve the surface fluid dynamics of aircraft, ship hulls, and wind turbines. Dr. Volkmar Stenzel, Yvonne Wilke, and Manfred Peschka of the Fraunhofer Institute for Manufacturing Engineering and Applied Materials Research created a paint that significantly reduces drag on aircraft. For their work, in 2010 the team won the prestigious Joseph von Fraunhofer Prize from Fraunhofer-Gesellschaft, Europe's largest research organization for applied sciences.

The paint also uses microscopic nanoparticles (which are biomimetic in their own way by copying nature's strategy for assembling things one molecule at a time) in its formulation to withstand UV radiation and extreme temperature changes, such as those found at high altitudes and endured by aircraft. "This paint offers more advantages," explains Dr. Stenzel in the institute's Research News Special Edition May 2010 report. "It is applied as the outermost coating on the plane, so that no other layer of material is required. It adds no additional weight and dramatically reduces the cost of stripping a plane every five years when all the paint is removed and then reapplied. In addition, it can be adhered to complex three-dimensional surfaces without a problem."

As well as designing the paint's molecular structure, the team developed its method of application. "Our solution consisted of not applying the paint directly, but instead through a stencil," explains Manfred Peschka. "This gives the paint its sharkskin structure. The unique challenge was to apply the fluid paint evenly in a thin layer on the stencil, and at the same time ensure that it can again be detached from the base even after UV radiation, which is required for hardening."

This technology can be used to great effect on fuselages. The German researchers were "able to reduce hull friction by more than 5 percent in a test conducted with a shipbuilding facility." Translated into increased fuel efficiency, such an improvement can represent an enormous benefit to shipowners. "Extrapolated over just one year, it would mean a potential savings of two thousand tons of fuel for a large container ship" typically traveling thousands of nautical miles. If applied to the world's aircraft, the total savings could be almost four and a half million tons of fuel per year.

The paint's developers at the Fraunhofer Institute claim that it will cost no more than conventional products to deploy. Such a technology has the potential to dominate both the lucrative marine and aerospace paint industries. How lucrative? As one example, the new cruise ship *Oasis of the Seas* used 166,000 gallons of antifouling material during its 2010 dry dock visit—and there are fifty thousand large ships in the world. One thin layer of this easily applied, nontoxic "shark" paint could save not only billions in fuel costs but also eliminate the carbon dioxide emissions resulting from burning all that extra fuel. This is an excellent example of the huge economic and environmental benefits just waiting to be gained through biomimicry.

Fast-moving fish, like tuna, rarely have creatures like algae, worms, or barnacles attached to them. Slower moving animals, such as turtles and some whales, do attract their share of these freeloaders, which start with the omnipresent oceanic populations of tiny microorganisms or bacteria attaching to the animal's outer skin. Despite being known for their speed (it's reported that a mako can hit 60 miles per hour when attacking), sharks actually spend most their time cruising in low gear to conserve energy. One would think they'd attract their share of these microscopic creatures. It turns out, though, that the shark's dermal denticles—the same feature of their skin that produces less drag—also has the added benefit of attracting fewer surface passengers. Which is exactly what inspired Sharklet Technologies, based in Colorado, to develop a thin film that mimics these denticles and resists the intrepid colonization of microorganisms on water-bound surfaces.

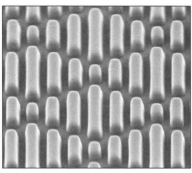

Sharklet surface

Dr. Anthony Brennan and his team in the University of Florida developed the Sharklet technology through a research program funded by the U.S. Office of Naval Research. As the company's website explains, the challenge was to develop a nontoxic way to prevent microorganisms, and subsequently barnacles, from attaching to ship's hulls. Dr. Brennan, an engineering and materials science professor, is an expert in the life cycle of these tiny creatures. As many microorganisms "can grow and divide every twenty minutes, a single cell can become more than eight billion cells in less than twenty-four hours." Once these cells attach to a surface, they rapidly colonize into so-called biofilms, which then lead to algae formation, weed growth, and other forms of fouling on the hull. Ship operators know that dragging all these extra passengers through the water has a negative impact on fuel use. A vessel can consume up to 40 percent more fuel if it is fouled, so whether it's a navy aircraft carrier, cargo vessel, cruise ship, fishing vessel, ferry, or small sailboat, boat owners around the world must regularly haul their boats out of the water to clean underwater surfaces. These annual or biannual dry dock sessions are not only expensive in labor and materials but represent nonproductive downtime for the boat owner. This is an even more critical concern for military ships.

In the past, a variety of toxic paints were applied to kill or resist microorganisms, but the harmful environmental effects from these is so severe that the most effective paints are now banned. Dr. Brennan's breakthrough came as he and his colleagues watched a submarine with a fouled hull return to Pearl Harbor in Oahu. He commented that it looked like an algae-covered whale, and asked himself and his team if there were any slow-moving sea animals that weren't victim to the same fate. Sharks captured his attention.

In development since winning its first research grant in 2002, initial commercial products of Dr. Brennan's research entered the market in 2007. The Sharklet surface is composed of millions of microscopic, diamond-shaped denticles. Each is 1/5th the thickness of a human hair wide and 1/40th deep, so the pattern can't be seen with the naked eye. Like shark's skin, each denticle has a particular height-to-width ratio, which creates just the right amount of surface roughness to discourage microorganisms from attaching and colonizing. Just as we wouldn't find it comfortable to lie or walk on a bed of jagged rocks, it takes too much energy for bacteria to hold on and reproduce efficiently, so they slide off to an easier place (or species) on which to perch and multiply. The spacing and height of tiny ribs on sharks' denticles, mirrored in Sharklet's technology, also interfere with the bacteria's ability to signal one another. Performance results are compelling. For example, the company describes that "green algae settlement on the Sharklet surface is reduced by 85 percent compared to smooth surfaces. Sharklet is now the first nontoxic, long-lasting, no-kill surface proven to control the growth of bacteria." With that valuable feature, there are obviously many commercial opportunities for the technology in marine applications. Surprisingly, this biomimetic technology can also be used in the multibillion-dollar health care industry, since the same resistance to microorganisms can be applied to prevent bacteria from latching on and colonizing objects in medical settings. After analysis, Sharklet started by selling a range of low-cost biomedical products that could be easily installed by consumers. Sharklet's adhesive-backed plastic films can be purchased directly from their website and attached to door plates, bathroom stalls, bed rails, trays, locker room benches, and other custom sites to reduce the spread of infection. After positive response to their first products, the company is developing other applications, including recently being awarded a grant by the National Institutes of Health to develop a catheter to diminish the rate of urinary tract infections in hospitals.

WRAPPED FOR FUEL EFFICIENCY

SkinzWraps, Inc., based in Dallas, Texas, is marketing a variation on the sharkskin theme to improve energy efficiency in vehicles. The company, founded in 2001, produces sheets of very thin, brightly colored plastic printed with designs and logos, which stick onto the surface of cars, buses, or trucks for marketing purposes. SkinzWraps has wrapped more than seven thousand vehicles to date. You may have even seen their wraps yourself—on city buses advertising hotels, Volkswagen Bugs turned into rolling ads for insurance, or trailer trucks promoting pizza parlors. SkinzWraps' biomimetic leap came with the notion of modifying their advertising material to enhance the aerodynamics of a moving vehicle. The company's newest product, the patent-pending Fastskinz MPG-Plus wrap, was launched in 2008. It has dimples that use the same underlying strategy as sharkskin, to improve the interchange at the boundary between the surface and the surrounding air. SkinzWraps claims that users can save up to 20 percent of fuel consumption in conventional vehicles and up to 25 percent of fuel in aerodynamically conscious hybrid vehicles.

Adapting biomimicry into an existing small business, as SkinzWraps has done, produces several benefits. Technology, whether biomimetic or otherwise, often transfers slowly from a university research lab to either an entrepreneurial start-up company formed for the purpose or a public company with multiple departments. These represent two ends of the risk spectrum: The entrepreneurial company may be passionate but not market savvy or well funded, while the large corporation may understand and even own a significant part of the market but be driven by short-term profit margins and subject to projects being cut at the stroke of a corporate pen. Instead, many of the functions necessary to rapidly exploit a biomimetic concept are already in place in a small business. SkinzWraps, for instance, had personnel who understood the critical qualities of their base product, including its manufacture, distribution, and marketing. While at its start it was an innovative advertising company, SkinzWraps married its technology to a biomimetic advertising model to provide a compelling new case of fuel savings to its clients. This also greatly expanded their target market to include the general public.

Sharkskin doesn't just optimize fuel efficiency in products that rely on gasoline and other fuels but also can have applications in the world of human fuel and energy efficiency. Take, for example, Speedo—the leading designer of racing swimsuits that has successfully wrapped humans in simulated sharkskin. The popular company's Fastskin helped athletes win big in the 2004 Olympics, and their low-drag LZR suit was so effective in the 2008 Beijing Olympics that it was worn by twenty-three of the twenty-five swimmers who set new world records. These high-tech suits, which cost up to $600, were designed using state-of-the-art computer simulation models, NASA wind tunnels, and input from the Australian Institute of Sport. Made of a patented textile that has tiny triangular projections, like sharkskin, to help stream water off the racer, the suits gain speed from several other features, as well. As described by Textile World.com, some of the seams are now "glued" ultrasonically rather than stitched, which means no water can penetrate between the stitches. This creates a water-impermeable barrier like a seal's skin. Built-in core stabilizer and internal compression panels—somewhat like corsets or girdles woven into the fabric—stabilize the swimmers' muscles and position them best along the surface as they move through the water. Whether or not the swimwear designers intended it, the result presents a similar profile to the water as marine mammals. The international organization for competitive swimming, FINA, was sufficiently struck by what seemed to be an unfair advantage offered by Speedo's technology that in 2009, it set a number of restrictions on the use of such slippery suits. Rather than covering the whole body, men's suits can't cover more than from waist to knees, and women's suits must cover no more than shoulders to knees. Perhaps fearing that even more biomimetic innovation might be around the corner (shark fins, perhaps?), FINA also ruled that suits must be made entirely of textile, or woven material, without any attachments or fasteners.

Computer simulation of water flow over swimmer,
with colors showing differences in areas of drag

PADDLES OF POWER

In another shark-inspired technology, the Australian company BioPower Systems has adapted the efficiencies of a shark's tail fin to improve the process of generating electricity from the movement of water. Humans have used water turbines for hundreds of years. Mills, often situated on creeks or rivers, used the force of water to turn wooden paddles. They, in turn, rotated one heavy stone on top of another to grind wheat into flour. Once electricity became widespread, the same basic concept was used to generate electricity. In a hydroelectric power plant, the force of water falling from a dam activates a turbine impeller—a rotating set of blades somewhat like a boat propeller. The turbine is attached to a shaft, which then rotates in a magnet in a coil, generating electricity.

Today, hydroelectric plants account for 16 percent of the world's electricity generation, but while the global demand for electricity keeps increasing, new rivers to dam are hard to come by. Attention is now turning to the oceans. Some parts of the world have huge tidal ranges, up to a staggering fifty-five-foot rise and fall in sea level every six hours. This results in the flowing exchange of billions of tons of water, in or out, four times a day. By installing an impeller or blade in the water, the force of passing tides, waves, or in some cases even strong currents, can generate clean electricity. In fact, it has been calculated that exploitable tidal power in the United States could provide a significant amount of the nation's electrical needs.

Ocean hydropower is complex from an engineering standpoint, however. The devices need to not only send power back to shore through heavy submarine cables but also be sturdy enough to survive storms and corrosive salt, avoid causing damage to or being damaged by boats, resist fouling by weeds and barnacles, be maintained and repaired while underwater, and not harm the environment or marine life. There are political issues at play as well. Who owns the water beyond the shore? Is it the local city or county? Or the national government that patrols it with its coast guard? It's like mining rights, where certain privileges belong to the local owner and others to the state or federal government. How far offshore do their various claims extend? Despite the complexities, it's a promising and emerging field. Although there are only a few dozen tidal or wave power plants currently operating in the world, a number of governments, universities, and start-up companies have recognized the opportunity and are actively conducting research and developing pilot sites.

BioPower System's bioSTREAM project is one example of novel hydro-turbine technology. Its biomimetic inspiration was helped into action by research on the extraordinary efficiency of tail fin propulsion in sharks and tuna. Brooke Flammang of Harvard University found that a shark creates whirlpools, or vortices of water, by alternatively flexing and stiffening its tail muscles. The shark can glide forward more efficiently on these watery ball bearings. After extensively analyzing fish propulsion, researchers at Massachusetts Institute of Technology and colleagues at Woods Hole Oceanographic Institution have built robotic tunas that demonstrate propulsion efficiencies of more than 85 percent. That's considerably better than typical ship propellers, which are often less than 50 percent efficient. Even the finest props struggle to achieve 70 percent.

BioPower applied the principles of tail fin propulsion in an inverted way. Instead of creating something that swims through waves more efficiently, the company plans to build devices that generate between 250 and 1,000 kilowatts of power by letting the waves themselves swim the tail-like paddles of their novel hydropower generators. Their pilot plant is a 250-kilowatt installation with an array of tails that are approximately 50 feet high and 66 feet long. Looking like a supersophisticated weather vane, each tail has an onboard computer that monitors the strength of the current (in this case water instead of wind) and can automatically adjust the angle of the tail to relieve pressure if the current gets too strong. This ability to reduce damge from buffeting means the construction can be less heavy-duty and therefore less expensive to build than competitor hydroturbine blades. BioPower recently received over $10 million in grants from the Australian state of Victoria and the federal government to support their demonstration project.

Biopower's shark tail hydroturbine array

About a year after that adventure in the Sholl Islands, I found myself on another expedition aboard a patrol boat off the west coast of Australia in the Abrolhos Islands, which consist of a chain of 122 islands in the Indian Ocean. And it was there that I had a memorable close encounter with a humpback whale. As I scrambled up from below decks at 6 a.m. to get ready for another day of seafaring, my skipper, Ted, exclaimed, "You should see the size of this whale swimming out from under the boat!"

"You should see its tail and calf," I yelled back as I watched two, very large shadows pass under the boat. The scene was fitting for the Abrolhos Islands. *Abrolhos* means "look out!" and was named by early Dutch navigators who crashed their lumbering sailing ships into them when inaccurately plotting their course to the Spice Islands north of Australia. The Abrolhos is comprised of tens of thousands of acres of coral reefs and lagoons, capped by tiny, rocky outcrops that barely qualify as islands. The water in our anchorage lagoon was about four acres in size, sixty feet deep, and crystal clear, with up to an astounding 180 feet visibility on some days.

A humpback whale had entered the lagoon, probably to give birth or to defend her calf. The water was clear and warm. To swim with a whale in these conditions was the opportunity of a lifetime. I had to go in, though I felt at some risk, knowing that such a massive animal could easily dispatch me in one swift movement. With my heart in my mouth, I donned my mask and fins. I slipped over the side of the vessel, making as little splash as possible. I had entered the water facing the whales. As the bubbles cleared from my face mask, I could clearly see the mother whale, like a bus gliding by about fifty feet away, but at first no sign of the calf. I then caught a glimpse of the younger animal hiding behind its mother. She turned and headed toward me.

My heart pounded hard. The sheer size of my companion was gripping. About fifteen feet from me, she changed direction and turned side-on, staring at me as she cruised silently by. I was mesmerized. I felt completely exposed and simultaneously held by the dark depths of her huge eye. Although whales have an extensive language of song and, due to the noncompressible nature of water, can communicate with their relatives and potential mates up to five thousand miles away, this protective mother was in stealth mode. As she passed, I could see isolated clusters of white and gray barnacles clinging to her skin. The leading edge of her long flipper, looking more like an aircraft wing, had large, irregular bumps or growths along its length. They, and the barnacles, looked strangely out of place on her otherwise streamlined body.

MAKE NO MISTAKE

Through a surprisingly and somewhat humbling experience, university researcher Dr. Frank Fish discovered that these odd looking design "mistakes" on a whale's flippers are actually nature's ingenious strategy to not only reduce drag but to give her extraordinary maneuverability.

As a fluid dynamics researcher, Frank knew that irregularities on the trailing, or back, edge of a wing could reduce drag and noise. As his company's website describes, he saw a sculpture of a humpback whale in a store one day with bumps on the front of its flippers and made the observation: "Look at that. The sculptor put the bumps on the wrong side of the flipper." But the store manager knew the sculptor's work well enough to know the sculptor wouldn't have been wrong, especially when it came to humpbacks. So she assured him that the bumps were where they should be. After checking the facts, Frank conceded that the artist was indeed correct. But if the artist was right, then contemporary scientific understanding was wrong. It led Frank to dig further, and now, in his Liquid Life Lab at West Chester University in Pennsylvania, he's a specialist on the fins of cetaceans.

Cetaceans, from the Latin *cetus*, which means "whale," are a class of marine mammals that include whales, dolphins, and porpoises. Frank Fish is now one of the world's leading expert on how their shapes result in phenomenal hydrodynamics. If a plane climbs or turns too steeply, the flow of air can separate from the wing as it rushes past. This makes the plane lose the lift of its wings so that it stalls (the technical term). Proper flight requires keeping flow attached to the wing, since this dangerous stall condition can result in a plane falling out of the sky.

Humpback whales need to climb and turn rapidly to first herd and then gulp balls of small, schooling fish. Dr. Fish's research confirmed that the irregular bumps on the whales' fins act like gutters or channels that guide the water flowing past into many tiny, spiraling whirlpools. These whirlpools, or vortices, are made of the same water that surrounds them, but the way the water rotates does two things for the whale. First, the vortices reduce drag on the whale in a similar way that a shark's denticles corral water flow. The bumps help create a thin cushion of tiny, spinning whirlpools that act a little like ball bearings close to the animal's skin, so that the bulk of the water can slide past. Second, these little whirlpools each lower the pressure within them so that the water flowing past is sucked closer to the whale and is less likely to separate when rushing past the fin. This helps the animal maintain lift and avoid stall even when maneuvering at steep angles. When compared to a regularly shaped wing or blade that you might find on any airplane or fan, Dr. Fish's research has shown that a whale's bumpy fin can provide 8 percent increase in lift and a 40 percent increase in its angle of attack, enhancing its ability to fly and turn the animal through its watery atmosphere. To engineers, whose designs benefit from one hundred years of aeronautic analysis and innovation, a 40 percent improvement over their work is a staggering amount. The bumps also provide around 30 percent reduction in drag, to better slice the fins through the surrounding water.

Dr. Fish proved once again that nature can confound our traditional scientific logic and produce performance far beyond human engineering imagination. His work is now patented and is being commercialized by WhalePower, a company based in Toronto, Canada, that's applying whale bumps to the leading edge of wind turbines and large fans in order to improve their energy efficiency and reduce noise.

Founded in 2006, the company is completing the onerous but necessary multiyear testing of wind-turbine blades. Their technology was considered innovative enough to receive research funding from the Ontario Centres of Excellence. WhalePower reports that performance results exceeded the company's engineering predictions, including the way that their blades responded smoothly to sudden gusts of wind, which would usually disrupt or damage a conventional blade. When a typical blade sees high variation and gusts of wind speeds, it can go in and out of stall conditions, with the airflow separating as it passes the blade. Like a whale using its fins effectively to steeply curve around as it swims, WhalePower's approach of adding bumps to the front of the turbine blades helps minimize the impact of variations in wind speed. The channeling of air by the bumps results in less strain on the equipment and expands the range of wind speeds that can generate power, even in gusty conditions. WhalePower has flown (wind industry language) a ten-meter prototype blade for six months that was independently tested by the Wind Energy Institute of Canada. While exact results are confidential due to WhalePower's commercialization process, the blades showed novel performance in response to wind gusts and low wind speeds as well as reduction of tip chatter and noise.

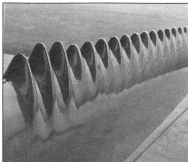

Whale flipper and WhalePower blade

Stephen Dewar, vice president of operations, cofounded WhalePower with Frank Fish, Bill Bateman and Phil Watts. I've known Stephen for several years and have watched with interest the challenges faced by the company since it was incorporated. I recently asked him if he had advice for aspiring biomimicry entrepreneurs. His number one message: Have a clear vision of why you're taking on your project and where you want to go. Then, like nature, be flexible and adaptable. His staff even made T-shirts that quote one of his favorite sayings: "For every vision, there is an equal and opposite revision." Second: Do your homework. Be deeply interested in the science, so you can clearly differentiate what's not just a good biomimetic idea but one for which there's strong market demand.

Stephen started his working career as a radio producer with the Canadian Broadcasting Corporation. He went on to produce successful TV shows and then founded an optical disk company, Paralight Laser Technologies. "It was there I learned not to trust the bandits at megacorporations," says Stephen. "They broke every deal."

How did a radio and TV producer and optical disk manufacturer become committed to biomimicry and whale bumps on turbines?

"I was fascinated by the physics of biological systems," he told me. "It's incredible how nature keeps standing conventional thinking on its head. When I heard about aircraft wings possibly benefiting from the design of whale tubercles, I tracked down Frank Fish and his colleague, Laurens Howe, and the rest is history. Given the safety issues and regulations involved with aviation, it was clear to me that adding whale bumps to the wings of aircraft would take a decade or two to get to market, so I suggested we look at rotating blades like wind turbines. We got a grant to retrofit an existing system and saw a 20 percent improvement in annual electrical production."

Commercial interest came quickly for WhalePower, and the company signed nondisclosure confidentiality agreements with three of the largest wind turbine manufacturers in the world. Then the great recession hit and deals that were almost in place died on the vine.

Stephen went on to say, "Western plans for wind farms and even existing turbine plants were mothballed. The big players went to China, where they found the government willing to fund new fields. With little credit available, fewer wind fields have been built and existing players are reluctant to invest in R&D or to change tooling." Stephen continued, "When there are no corporate dollars and no government dollars, how does a nation grow its new technologies? I'm just staggered at the austerity and extreme thinking happening in the United States. Austerity is mutual suicide. It's depressing and flawed thinking."

WhalePower is a closely held corporation, backed by a small group of investors. It keeps its business model focused solely on licensing. Rather than building a large team, the company keeps its overheads low and uses outside contractors to meet the needs of particular projects. To commercialize their technology without building large infrastructure, the company has partnered with a fan manufacturing company, Envira-North Systems Ltd., to build and sell a line of high-volume, low-speed large fans for use in agricultural and industrial sites. The whale bumps on the front of the blades reduce drag and increase stability and efficiency. This and the way the air is channeled past the bumps reduce the loud whirring noise and chatter created by most high-speed fans, while still producing the same amount of air movement.

WhalePower deliberately chose product applications where their customers, such as the operators of farms with large agricultural sheds that need ventilation year-round, will directly benefit from the energy savings of a more-efficient fan. Along with the money saved on electricity bills, which provides payback of the fan purchase and installation costs, the reduced noise that comes with WhalePower's natural design is a welcome bonus to nearby animals and people.

Fans consume more than 20 percent of the electricity used on earth, so they present an important opportunity for innovation. Approaching the same industry from a different biomimetic inspiration, one of our companies, PaxFan, improves efficiency by integrating nature's three-dimensional spiraling geometry, especially in smaller fans where leading-edge bumps have less effect. Both companies are adapting nature's use of vortices and are exploring ways they might combine their technologies to create even more effective solutions.

A WHALE OF GOOD IDEAS

Cetaceans are a group of mammals that returned to the oceans after living on land. Their flippers are actually adapted front legs, and a whale's skeleton clearly shows the fingers of its ancient land ancestor.

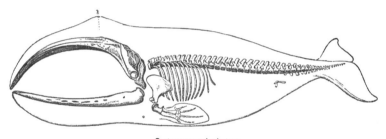

Cetacean skeleton

How these large land mammals evolved into sea creatures has fascinated and puzzled scientists for centuries, though recent findings of fifty-million-year-old fossils in Pakistan have filled key gaps in the story. Scientists knew that cetaceans started as land dwellers, because they have lungs and are air-breathers. What's more, their bones follow the same lines as land animals' limbs, and their backbones move like a mammal running instead of a fish swimming. But what did they descend from? For decades, their ancestors were thought to be distant relatives of camels, with hooves and pointed teeth, because some whales have teeth that match the pointed teeth and hearing structure of an ancient land animal that somewhat resembles an imagined cross between a camel and a wolf. Several years ago, researchers discovered missing link fossils, proving that whales, dolphins, and porpoises actually have the same ancient ancestors as the hippopotamus. It makes sense: Hippopotamuses are heavy mammals that spend much of their lives in water. Millions of years ago, a branch of the family evolved to be heavier and heavier as they spent more time in weight-supporting water. Over millions of years they adapted into the graceful swimmers they are today.

Humans have been in the business of whale hunting for thousands of years and, by the 1960s, had almost wiped out many species. Thanks to conservation efforts worldwide, however, the total population of whales in the world's oceans is increasing, and the hunt for biomimetic solutions based on whales is gaining speed as well. Whales range in size from the 115-foot, 150-ton blue whale, the largest animal ever to have existed on earth, down to the pygmy sperm whale at a diminutive eleven feet in length. Interestingly, the easily recognized black-and-white killer whales are not whales at all but are actually orcas, the largest species of dolphin.

There are two types of whales: toothed whales and filter feeders. The latter take huge gulps of small schooling fish or plankton-rich seawater and strain it back into the surrounding sea by squeezing it through a comblike mat of stiff baleen, leaving a bouillabaisse dinner behind in their mouths.

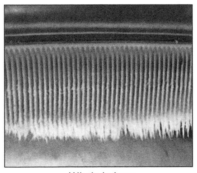
Whale baleen

Imagine straining a mouthful of soup by gripping your teeth together and parting your lips, using your tongue to squeeze out the broth, keeping the meat and vegetables inside the fence of your teeth. Whales need to clean their mouths of leftover crumbs, so they have also evolved a type of flush-out process that happens when they dive and then surface for their next gulp.

Baleen is a curious material made from keratin—similar to the fingernails, hair, and hooves of other mammals. It was marketed as whalebone and was essentially the plastic of the late 1800s, being used for corset stays, hoop skirts, umbrella ribs, fishing rods, buggy whip handles, and carriage springs.

Now, thanks to research by a team at the University of South Australia, baleen has inspired a patented separation technology called the Baleen Filter. Without adding any chemicals, the Baleen Filter offers filtration of liquids down to 25 microns. A micron is 1000th of a millimeter; by comparison, a human hair is between 50 and 150 microns and the smallest thing that can be seen with the naked eye is about 10 microns, so 25-micron filtering without any chemical help is highly effective. With chemical assistance, the filter can achieve filtration down to 3 microns. Like a baleen whale squeezing seawater out of its mouth and leaving its lunch behind to be swallowed, contaminated water is washed through a Baleen Filter, and then, mimicking how the whale's tongue sweeps across baleen, trapped solids are sluiced from its surface and flushed into a reservoir (instead of the whale's stomach) for future handling. There are only two outputs of the Baleen Filter—filtered liquids and separated solids.

Based in Adelaide, South Australia, Baleen Filters Pty Ltd. was founded in 1999 after four years of research. It currently distributes its products, which are not only highly efficient, but also self-cleaning, to the fish processing, cured meats, pig farming, and other food and agricultural industries. Potential applications for the technology include "simplified wastewater handling practices for environmentally sustainable development projects, and by-product opportunities for recovered solids, including biogas generation, soil conditioners, vermiculture (worm composting), farm feeds, tallow, and protein recovery." With its ability to separate liquids from solids in a portable, nontoxic way, the company also sees opportunities in mobile cleanup systems and emergency pollution response.

Like all animals that have survived on earth for millions of years, every species of whale represents an extraordinary set of solutions to life's challenges. Each of these can be studied and adapted, making a single whale an entire library of design expertise and information. For example, the filter-feeding gray whale has an enormous tongue, weighing up to three thousand pounds, that makes up to 5 percent of its total body surface area—the human equivalent would mean that a two hundred pound man had a ten-pound tongue! Researchers John Heyning and James Mead determined that the whale's tongue acts as a huge heat exchanger, a device that transfers temperature from one medium to another. Heat exchangers are used in a wide range of industries as a key part of any heating or cooling system, including space heaters, car radiators, refrigerators, and air conditioners.

All baleen or filter-feeding whales need to pass very large amounts of cold water loaded with plankton through their cavernous mouths. Although a whale's warm-blooded body is insulated with blubber that can be up to eighteen inches thick, its tongue doesn't have such a layer of protection. Nature can't afford the whale to lose essential body heat every time it gulps a mouthful of ocean water, so it has evolved the tongue into something like a giant, fleshy radiator. The whale achieves this with a complex network of blood vessels inside its tongue. These are laid out in such a way as to retain heat while "precooling" the blood that gets sent to the tongue's surface, so that it's not shocked by the cold water. So effective is this process that "the surface temperature of the tongue of a young gray whale was measured to be only half a degree Celsius higher than the (surrounding) water," while the whale still maintained its overall body temperature. Scientists are researching this strategy for application in computer and electronic cooling and heat exchange in aircraft, buildings, or power stations.

Although there's an old saying, "If you don't have your health, you don't have anything," the escalating cost of health care has become an economic and political football in the United States. Heart pacemakers have helped millions of people worldwide since their invention. As a summary for the United Nations Environment Program describes, at up to $50,000 for each procedure, and around one third of a million operations per year just in the United States, pacemakers add up to a global expenditure of $3.7 billion per year. The human heart weighs ten ounces and pumps blood through sixty thousand miles of veins and arteries. The humpback whale's two-thousand-pound heart effectively pumps enough blood to fill a small swimming pool, through forty-five hundred times as many veins and arteries as humans—with as few as "three or four beats a minute" at times. Through living echocardiographs and autopsies on dead whales, Jorge Reynolds and his team at the Whale Heart Satellite Tracking Program in Columbia discovered microscopic wire-like structures in whale hearts "that allow electrical signals to stimulate heartbeats even through masses of nonconductive blubber." They believe the technology can be adapted to human pacemakers with a new version that would cost a fraction of the cost to make, use no batteries, and need far fewer follow-up procedures. The potential savings to national health budgets and patient health is compelling.

It was November 1985, and a pitch-black, steamy night in the Java Sea. En route to Singapore from Australia, our fifty-foot sailboat should have been lit up by at least the tricolor navigation light at the top of her mast. Instead, the night before we had been hit by lightning, which shorted out all the boat's electrical systems. There was no wind, so the main engine was propelling us and I had rigged a small flashlight, covered with red cellophane, to the compass so we could see to steer by. About midnight, I looked behind us and was enraptured by a broad trail of exquisite sparkling lights stretching back a mile or two from the stern of the boat. Millions of fluorescing organisms had been scintillated by the turbulent rotations, or cavitation, of our propeller. It looked like a scene from a Peter Pan fantasy—one of the most beautiful and grand sights I have ever seen.

Bioluminescence means "living light," and that's exactly what it is. While specific mechanisms differ across the many species that use bioluminescence as part of their survival or mating strategies, it always emits from a living organism and involves the oxidation of an organic molecule. By dramatically agitating the water with our churning propeller, we were, in effect, turning on their chemical lighting systems.

A few minutes later I heard the familiar whoosh of dolphins expelling air as they surfaced. I rushed up the deck to stand on the cathead, a small platform protruding from the bow. When you're standing on it, you feel like you're suspended in the air over the ocean. It's an enthralling experience, especially when dolphins are swimming and breaching through the surface just six feet below your feet.

We were often accompanied by dolphins. As many as forty or fifty in a pod swam with us, in and out of our bow wave, sometimes for fifteen minutes or more. This night, as they leaped out of the water, they trailed cascades of sparkles. As they reentered the water, they pulled air and sparkles with them, clearly defining their tracks underwater. I was surprised to observe that after a few seconds below the surface, the dolphins created no more fluorescence until they broke the surface again. As I watched them play, admiring the fact that the dolphins did not disturb the fluorescing organisms, no matter how fast they swam, I began to wonder: Could traditional boat designs and propellers be biomimetically re-created to achieve such efficient motion?

Turbulence, or cavitation, is a negative consequence of boat propellers. It results from the rotating blades disturbing the water so vigorously that it essentially boils into millions of tiny bubbles of vapor. When the bubbles collapse, it can be with a force of as much as twenty thousand pounds of pressure per square inch. (A car tire is typically inflated to just thirty-five pounds per square inch. A can of Coca-Cola has about forty-five pounds of pressure per square inch.) In fact, these explosions are so violent that they tear chunks of metal away from the propellers and can eventually destroy them. I've heard of a ship, that, on its maiden voyage from the United States to England, lost blades from its $100,000 propeller—in just one trip.

The bioluminescence I saw that night causes the world's navies considerable angst. Modern stealth technology has made naval ships almost invisible to radar. What they can't control in waters with bioluminescent organisms, however, are the miles-long light shows trailing their vessels at night that are clearly visible to enemy satellites.

In the years that followed that sailing trip, my efforts to create a propeller that wouldn't produce the turbulence I had seen scintillating the

water in the Java Sea led me to design a whole series of devices that move
water, air, and other fluids. I also designed two series of boats modeled
on the dolphins' ultrastreamlined shape: the WildThing and the XAP.
Nimble and stable, these watercraft have been used in a wide variety
of applications, from search and rescue to fishing to diving to racing.
Their hull designs copy the way dolphins and other water animals cut
drag. The ratios of their pontoons mirror the body shape of dolphins
and porpoises.

From left to right: cetacean inspiration, WildThing watercraft

The egglike, compound curves of the boat hulls make them strong without needing heavy internal support, so they're a fraction of the weight of
similar craft. Coupled with streamlined design, this translates into a third
less fuel consumed than with conventional approaches. In the early 1990s,
we built and sold several thousand WildThings and then partnered with
an Australian venture capital firm to expand into global manufacturing
and distribution. Unfortunately, the company was forced to shut down a
few years later after its controlling venture capital firm was investigated by
federal authorities for fraudulent handling of various investments.

Nature's fastest objects, almost invariably, have flexible or adaptable
surfaces, but boats, like most human-made technologies, generally have
hard skins of wood, fiberglass, or metal. My latest boat design, the XAP,
also adapts a particular characteristic of a dolphin's skin. When the animal needs to lower drag to put on a burst of speed, its skin develops goose
bumps, somewhat like the dimples in a golf ball. Similarly, the XAP's
surface isn't shiny slick, but is instead slightly textured in a biomimetic
drag-reducing pattern.

XAP watercraft

THE DOLPHIN NETWORK

Dolphins, and their smaller cousins, the porpoises, are universally revered marine mammals. Ranging from the four-foot-long, Maui dolphin, to the thirty-foot orca, the forty species of dolphin and six species of porpoise appear as positive symbols in the mythology of all cultures that are familiar with them. Ever wonder about the difference between dolphins and porpoises? Porpoises have flatter teeth and stubbier noses; this led to their name, which came from Medieval Latin *porcopiscis* meaning "pig fish."

Interestingly, dolphins and porpoises have not been generally seen as a food source and are eaten in very few places, such as Japan. As many as twenty-five thousand are hunted each year with estimates that hundreds of thousands more perish through by-catch of commercial fishing operations. Known for their extraordinary intelligence, playful nature, and willingness to interact with humans they're even attributed, by some folklore, with almost magical powers. Their swimming speed, antics, and endearing, chirping echolocation calls are enjoyed by swimmers, surfers, and divers worldwide. In fact, a dolphin's whistles, pulses, and clicks, made by air sacs just below its blowhole, are among the loudest noises made by marine animals. A scientist at Penn State's Center for Information and Communications Technology Research has been analyzing these underwater messages not for meaning but for hints on how to make our wireless signals more effective. As the Ask Nature database describes, Dr. Mohsen Kavehrad uses "multirate, ultrashort laser pulses, or wavelets, that mimic dolphin chirps, to make optical wireless signals that can better penetrate fog, clouds, and other adverse weather conditions." The multiburst quality of dolphin sounds "increases the chances that a signal will get past obstacles" in the surrounding water. In the same way, Dr. Kavehrad's simulated dolphin chirps increase the odds of getting around such tiny obstacles as droplets of fog or rain. This strategy could expand the capability of optical bandwidth to carry even greater amounts of information. Such an application technology could optimize communication between aircraft and military vehicles, hospital wards, school campus buildings, emergency response teams, and citywide networks.

Using a different aspect of dolphin sounds, EvoLogics, a company based in Germany, designs and manufactures "underwater acoustic modems developed after eight years of study on the physics of dolphin communication." Their "sweep-spread carrier technology delivers acoustic signals even in adverse underwater conditions," because it transmits with specific frequencies, based on dolphin calls, that persist over very long distances. EvoLogics reports that the devices, which can be applied for tasks ranging from guiding ships to monitoring areas for earthquakes, cost around $12,000 and have been installed in more than ten countries.

If you've ever seen dolphins swimming effortlessly around the bow of a boat or surfing waves it is clear to see how graceful, fast, and efficient they are. Some species of dolphin are able to maintain speeds of twenty-five miles an hour, hold their breaths for twenty minutes, and dive to depths of fourteen hundred feet. To watch delightful examples of dolphins that demonstrate uncanny mastery of advanced fluid dynamics, surf the Internet for key words like "dolphins play with bubbles" or "dolphins and vortex rings." You'll find videos that show them tossing and twirling silvery rings, which they make by blowing air out of their blowholes and pushing the resulting ribbons of air around with their heads and noses in what looks like playing with underwater Hula-Hoops.

Dolphins, like whales, need great maneuverability to hunt effectively. Unlike whales, however, dolphins have noses that can point at variable angles, giving them additional control over how their bodies interact with the surrounding water. Although the inspiration for the shape of the super-fast Concorde aircraft was not well documented when it was designed in the 1970s, engineers see a direct analogy between its innovative drooping nose profile and the way dolphins' noses improve their flight underwater and takeoff into playful leaps out of the water.

The dolphin's powerful tail inspired engineer and inventor Ted Ciamillo. He created the Lunocet, a two-and-a-half-pound, forty-two-inch-wide monofin. When an Olympic medalist like Michael Phelps is at his peak swimming performance, he converts only about 4 percent of the thrust and energy of his arms and legs into the forward motion of his body. A dolphin manages to convert 80 percent of its energy to thrust and forward movement. Wearing his biomimetic fin, Ciamillo doubles Phelps's top speed. Not only does his fin whet the appetites of competitive free divers, its benefits are not lost on the amphibious unit of the Marine Corps.

Another biomimic, Bob Evans, created compact diving fins, called ForceFins, after years of observation of dolphins and tuna. They provide strong thrust without fatiguing the diver and are used by amateurs and professionals around the world. These handsome swimming aids look quite organic and are even featured in New York's Museum of Modern Art. When tested by the U.S. Navy in a four-year study, it was found that divers used less oxygen while using Force Fins than the competing, bestselling, name brand fins. The two hundred divers who participated were asked to predict which fins would perform best and fatigue them the least. It's a testimony to the way our thinking can lead us astray that all the participants predicted that the stiffest and longest fin would give the best performance, when in reality, the smaller, much more flexible Force Fin proved the winner.

Until now, humans have seen the natural world as an inexhaustible resource. Whether whales or sharks, seals or otters, we essentially mined sea animals for their parts. Although we now use plastics and synthetic lubricants in place of whale baleen and oil, we continue to decimate whales and other marine life through harmful practices. Yet the animals and plants remaining on earth offer a truly inexhaustible resource to create a new global economy and virtually unlimited opportunities for wealth creation and problem solving. It's an entrepreneur's dream. Just the examples in this chapter from sharks and cetaceans, which are only two of hundreds of types of marine life, represent billions of dollars in new product potential and billions of gallons of saved fossil fuel. Like Yvonne Wilke, Frank Fish, Stephen Dewar, Mohsen Kavehrad, and Anthony Brennan, the scientists and businesspeople exploiting these new opportunities are having a whale of time, confident that they're part of a highly valuable and sustainable new field—a field many believe is the only viable future for humanity.

Chapter 5

....

SCALES AND FEATHERS

D amn! If it's not bad enough having to bounce up and down on these two wheel ruts, I have to go and break down in the middle of a creek crossing—in two-foot-deep water no less!"

It was 1976 and I was driving my government-issued Land Cruiser the hundred miles from the remote outback town of Broome to the cultured pearl farm at One Arm Point. This was hard country: blazing, subtropical sun and the soil bone-dry for nine months of the year, flooded for the other three, and sparsely vegetated. Attired in my standard khaki shorts and short-sleeved shirt, I took off my boots and socks and stepped out into the water. The engine had been sputtering and low on power for the past hour, so I assumed there must be crap in the carburetor. I carried spare fuel in long-range tanks, but in the vast, scarcely populated north of Australia, petrol, as we call it, can end up with all sorts of dirt, water, and other contaminants in it. I grabbed my tool kit, opened the hood, and dismantled the carburetor. Sure enough—flakes of rust in the jet and bowl. It took about an hour and a half, standing up to my knees in the muddy water, to complete the clean out and refit the parts. I closed the hood, threw the tool bag in the back, and climbed out of the water into the driver's seat.

Oh my God! That's when I noticed that my legs from the knees down were covered in large, glistening, black leeches. There must have been fifty or more, about the size of my little toe, on each leg. They were thoroughly attached and sucking my blood for all they were worth. Several had even chewed through the soft skin between my toes.

You don't actually feel leeches latching on or sucking. They anesthetize the area they pierce. Then they apply an anticoagulant to keep your blood flowing into their gut. They might not hurt, but they are seriously creepy. You can't pull them off. They're too slippery to get a grip on and they stick like chewing gum to a shoe—or shit to a blanket, as they say in Australia. So what do you do? First you must steady yourself and stop freaking out. Then apply one of the most satisfying solutions to a problem ever invented—salt. Just a tiny amount of salt placed on a leech's tail will cause it to instantly back out of you and fall off. I involuntarily shuddered as I grabbed a handful of salt from my food locker and wiped down my limbs—covering every square inch of my affected skin as rapidly as I could. Yuck, yuck, yuck! Thanks to the leeches' anticoagulant saliva, my legs were covered in blood. It looked like I had terrible injuries, but it appears much worse than it is. The salt stung, but it's one of those discomforts that one actually welcomes and feels good about—each sting meant that a leech had vacated.

Usually I think of all God's creatures as beautiful in their own way. The three exceptions are leeches, cockroaches, and centipedes. All right. They're all beautiful: masterpieces of design and function. I just don't want to share my personal space with them—ever.

WRIGGLERS

Leeches are cousins of earthworms. While some leeches eat small bugs or snails whole, most varieties feed by attaching to and sucking the blood of a larger animal before dropping off when full. Bloodletting, also known as hirudotherapy, by applying leeches to humans, has been used for thousands of years and is an example of biotherapy, where a living animal is used in the diagnosis or treatment of illness. Biotherapy ranges from dogs that are able to detect cancer or epilepsy to the use of bee venom to reduce joint pain from arthritis. Now biologists are studying the excretions and processes that allow a leech to attach to its host in the first place, without the host noticing, as well as the anticlotting chemical that leeches mix into their host's bloodstream to keep the blood from thickening and clogging their vampirelike slurping session.

Biomimetic, nontoxic, anticoagulants could spell big business in the treatment of heart disease and a number of other conditions. Blood clotting is an amazing process. When the body recognizes an opening or rupture in a blood cell wall, it sends chemical signals that cause a particular type of blood cell called a platelet to lay its disk-shaped form against the rupture. The contact of a platelet with the damaged cell wall results in a cascade of chemical reactions that converts fibrinogen, a soluble protein that floats around in blood to fibrin, an insoluble protein that forms a sticky mesh, or web. This closes the wound and stops the bleeding. If the body's chemistry is unbalanced, however, the chemicals that signal the blood to clot can trigger at the wrong time and cause medical crises, such as deep vein thrombosis, vascular blood clots, and embolisms (where a clot forms and travels to another part of the body). Anticoagulants such as warfarin, which is marketed under the brand name Coumadin, have saved many lives. Unfortunately, in part due to their disruption of the balance of vitamin K in the body, they come with a number of side effects—including nausea, headaches, loss of taste sensation—and potentially life threatening effects such as internal bleeding, bone instability, and artery calcification. Ironically, the same chemical, warfarin, is also one of the most widely used pesticides on earth. It kills rodents by causing them to bleed to death internally.

Though not inspired by a leech but by a paper cut, Dr. Ian McEwan applied the principle of blood platelets to an innovative purpose—the repair of pipes in water and oil industries. A specialist in particle-fluid transport problems, Ian founded Brinker Technology as a spinout from the University of Aberdeen in Scotland after realizing that our cuts heal by activating materials that are already present in the blood. Rather than having to stop production and drain a section of pipe to find and repair a leak, Brinker's Platelets are injected into the flow itself. When they get near a leak, the difference in pressure at the leak sucks the patented platelets against the hole to seal it. Brinker was named after the story of the little Dutch boy Hans Brinker, who saved his town by stopping a leak in a dyke. Unfortunately, the company filed for liquidation in early 2012.

A number of biomimicry researchers have also studied the neurons in leeches that control their movement and heartbeat rates for clues to optimized control systems, such as in heart pacemakers. The leech's heart is divided in two segments, one of which beats from front to back and the other from side to side. About every twenty beats, this pattern switches sides. Scientists are creating mathematical models to understand and replicate this seamless switching.

Albeit somewhat disconcerting for some to consider, back in the biotherapy realm, leech therapy has achieved FDA clearance and is experiencing resurgence in modern medical institutions, particularly in situations where the body is recovering from certain surgeries such as transplants of fingers or toes. During the healing process, blood flow is sometimes congested in constricted places and leeches can be used to remove blood that can't exit via the usual pathways.

"Nobody loves me, everybody hates me. Guess I'll go eat worms." So goes the children's song. Worms might not do much for depression, but Dr. Joel Weinstock, professor of medicine at Tufts Sackler School of Graduate Biomedical Sciences, discovered that swallowing the eggs of *Trichuris suis*, or whipworms found in pig intestines, can actually help control ulcerative colitis and Crohn's disease without the use of potentially dangerous steroids or cyclosporine,

Somewhat less controversial and slightly more palatable is the practice of eating earthworms as food. Dined on in almost ninety countries, earthworms are loaded with protein and minerals. These delectable wrigglers are best consumed steamed or smoked but are also sometimes eaten live and raw—a moving experience. Perhaps of more palatable importance to humanity is the adaptation of a particular worm's ooey-gooey-gluey technical genius: sea-worms' underwater house-building glue. Researchers at the University of Utah have synthesized the properties of sea worm adhesive, with an aim to create a nontoxic biodegradable superglue for mending bone fractures. Worm glue sets up in wet conditions, at ambient temperature, dissolves over time, and can be inoculated with and carry antibiotics, painkillers, or other medicines. Successful testing has already been conducted on cow bones.

Worm locomotion, rather than adhesion, is one area of study at the

Centre for Biomimetics at the University of Reading in the United Kingdom. Scientists there hope to design effective artificial muscles by adapting the way that worms worm their way around. It so happens that a worm's skin is a cloak of spiraling fibers that run the length of the creature. Like a tube made of interwoven string that can be stretched out to be long and thin or bunched up to be short and fat, these fibers allow Mr. Squiggle to expand and contract in three dimensions as he burrows through mud and earth—with extraordinary efficiency in biomechanical engineering terms. The researchers created pseudo worms by filling cylinders of helical fibers with a polymer gel that expands when water is added. By controlling the amount of water in the gel, the scientists can make the worm, or artificial muscle, expand and contract on command.

Whether we're dining or inventing biomimetic muscles, we all need light to see by. Earthworms could show us how to light the way through their bioluminescence. Environmental studies researchers at the University of Montana submitted a design proposal to the 2011 Biomimicry Design Challenge that lays out a compelling case for biomimetic lighting based on earthworms. Interacting chemicals, including a compound called luciferin and an enzyme called luciferase are both easily synthesized. Spontaneous reactions with hydrogen peroxide and water produce light and offer the potential for low-energy electronics and emergency lighting, signaling and night lights.

A rush of adrenaline left me on full alert. Apart from the pounding of my heart, it was otherwise a still and silent night in the farmhouse. The tortured screeching sound had lasted just the few seconds it took to wrench me out of a deep sleep. I lay there, barely breathing, unable to make sense of things. Thirty seconds passed and there it was again—that skin-crawling sound, followed by a metallic crack of something being struck against the iron roof. If you're familiar with the sounds of fingernails being dragged down a blackboard or a sheet of corrugated iron roofing, you can imagine the impact of waking to a greatly amplified version of it on the ceiling above your pillow.

Damn possums, I thought.

Australian and North American possums are both marsupials, but that's where the similarity ends. The American, ground-hugging, carnivorous opossum looks like a big rat, in part because of its hairless tail, and has more teeth than any other mammal. The Southern Hemisphere's deceptively cuddly looking, furry tailed, vegetarian animals live in trees and often colonize the ceilings of farmhouses. I guessed I'd heard claws sliding down the metal roof. The noise continued sporadically through the night. I'm going to have to fix it! Although often used to catch possums so they can be relocated, live traps are not so easy to deal with when loaded with a highly aggravated wild marsupial. An old-timer's remedy is to spread a couple of boxes of mothballs in the ceiling because the smell of naphthalene makes the area unattractive to animals. I bought six boxes to guarantee a good job and had them well distributed by lunchtime. That night, I went to bed reassured in the knowledge that I had solved my problems in a humane way.

Wrong. At one o'clock in the morning, I was woken by the sound of feet scurrying across the ceiling, back and forth, back and forth. It made no difference when I pounded on the ceiling with a broom. The sound of sharp nails on plaster, wood, and iron did not stop throughout the night. The next morning, I went outside, bleary-eyed, with a cup of tea to greet the day. To my surprise, I noticed mothballs strewn on the grass all around the house. My companion in the roof had obviously collected up the offending, odiferous balls and thrown them out from under the eaves, one by one. This couldn't be the work of a possum, could it? Armed with a flashlight, I gingerly put my head up through a two-foot by two-foot manhole in the ceiling. Sweeping my eyes rapidly left to right, fearful that I was going to feel the claws of an enraged possum in my scalp, I saw no sign of the animal.

But wait—what's that coming toward me—one large foot after another? Not a possum! I was staring at a large lizard, a good two feet long. These minidinosaurs have needle-sharp, long, curved claws and four strong legs to grip with. My friend was stepping relentlessly toward me and my light, which I hurriedly switched off. I descended from the manhole and regained my composure. I subsequently caught the beast by again enticing him toward the light—but this time throwing a blanket over him. I assume it was a him. I released Mr. Lizard about two miles away in a wilderness area.

SLITHERERS

There are nearly 3,800 species of lizards on earth, ranging in size from a little over an inch to the almost ten-foot-long Komodo dragon. In prehistory, water-dwelling mososaur lizards grew to sixty feet in length. Some extinctions paid off for humans.

Some lizards produce saliva that can kill you, or save your life. The oral juices of the giant Komodo dragon comprise a soup of fifty-two virulent bacteria varieties that completely overwhelm the immune system of anything unfortunate enough to receive even the slightest bite. Four of the bacteria species have no specific antidote and can kill you in as little as two days. Yet, if you're another Komodo dragon, you have complete immunity to a bite. Scientists are working to unlock this secret to create a whole new paradigm in biomimetic antibiotics. Molecular biologist Dr. Gill Diamond and his team at the New Jersey Dental School of the University of Medicine & Dentistry of New Jersey have already isolated two molecules in the dragon's blood that have particular antibiotic properties. His group is working to synthesize these molecules and start the lengthy process of developing new commercial drugs.

Parenti lizard

Another lizard, the two-foot long Gila monster of North America, eats very large meals—but infrequently. Nature has equipped it with a strategy for regulating its blood sugar between feasts. Its saliva contains a hormone that boosts insulin. Eli Lilly has synthesized the hormone into the drug, exenatide (trade name Byetta), a synthetic peptide that differs in chemical structure and pharmacological action from insulin, but has been successfully applied to the management of certain types of diabetes.

In a world of growing water scarcity, scientists are studying the water harvesting abilities of the Texas horned lizard and the thorny devil lizard. The former has a routine where he arches his abdomen, splays his legs, flattens his body, lowers his head, and slurps drinking water from rain that flows down and into his jaws. Awkward, but effective. The thorny devil gets his liquid refreshment from any water source including moist soil, against gravity and without a pump. Instead, he relies on capillary action to wick water up through tiny channels on his skin. By mimicking the lizard's micro-channels, engineers could use this strategy to provide cheap solutions for people in desperate need of water. Since it's a passive system, it requires little or no energy to harvest a drink and may even be applicable to the raising of water to the tops of tall buildings and to evaporative air-conditioning devices (also called swamp coolers).

Ball bearings are a $100-million market worldwide. They're used in a huge range of applications to reduce friction and increase mobility between moving parts. Another desert-dwelling lizard, the Sahara sandfish, swims through sand with extremely low friction, great abrasion resistance, and no ball bearings. This lizard slips through more than half a mile of dunes each day without showing any signs of wear. Evolutionary biologist Dr. Ingo Rechenberg of the Technical University of Berlin discovered that sandfish skin has 60 percent less friction than steel. He theorizes that the lizard's skin has silicon scale tips which could emit electrons and set up similar, repelling, negative magnetic fields in the sand grains. If this is true, our slippery friend is using a type of magnetic levitation as a friction-saving strategy. Its high performance scales, made from keratin (like human fingernails and hair) are giving researchers clues to the creation of a whole new range of materials that would need no lubrication.

Still and silent as a church at midnight, I was sitting and reading under the vaulted ceiling of a plantation-style house in Hawaii. I was enjoying my aloneness in the building, when suddenly—splat—something landed on my left shoulder. Startled out of my serenity, I looked up and saw the source of my blessings. I had become the target for a gecko that had relieved himself about fifteen feet above me. He was hanging upside down from the ceiling panels and I imagined that he had a satisfied look on his face. The following night, sitting in the same place with the same stillness . . . plop. Startled, I thought, no, not again, and turned my head to look at my shoulder. Instead of poop, there lay the gecko. My movement caused him to scurry down my arm and disappear behind the sofa cushions.

My gecko sighting reminded me of the dozens of research studies that have been conducted on how geckos are able to stick to and climb vertical glass windows and hang upside down from ceilings—at least, most of the time. They do this because they have a specially evolved skin surface on their feet that forms a high-energy bond to virtually any surface. Studies at the Max Planck Institute in Leipzig, Germany; the University of California, Berkeley; MIT; and many other institutions, including NSF-funded research in Ohio and at the Georgia Institute of Technology, have all contributed to the understanding of this gravity-defying ability. On each of a gecko's toes there are about a billion tiny, tubular hairs that are just nanometers long and wide. Each of these hairs is impacted by van der Waals forces that attract to a surface, so the gecko is in effect, sticky to start. In order to move, the animal simply relaxes a toe and the muscles operating the attachment mechanism let go. The same mechanism is also self-cleaning, which is hugely valuable. When the gecko toes let go, dirt is more attracted to the other surface he is walking on than to himself. He simply steps out of the dirt. Replication of these strategies is one of the most commonly cited examples of modern biomimicry. While commercial products are not yet widespread, development has resulted in chemical-free medical bandages and adhesives and could lead to the elimination of toxic or non-biodegradable glues for everything from Post-it type notes to wall hangings.

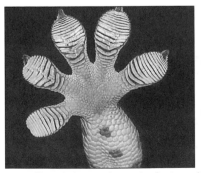

From left to right: gecko foot, gecko-inspired biodegradable bandage

If you think a gecko's sticky feet are cool, check out a chameleon's bug-catching, sticky tongue. Imagine if you were whacked in the head by something traveling at nearly thirty miles per hour. Researcher Alexis Debray, an engineer at Canon, has built a mimic of the chameleon tongue's remarkable acceleration—thirty-six feet per second—and its ability to retrieve an insect without creating such an impact that it sends it cartwheeling into space. Debray is developing his technology for rapid but highly controlled response in robots. An alternative use could be as an ultimate, high-tech, flyswatter—that won't actually squish the fly onto wall or table.

In June 1969, my skipper and I anchored the patrol vessel *Dampier* about three hundred yards offshore from the South Muiron Island, one of a pair of elongated, uninhabited islands off the northwest coast of Australia. Quite remote, isolated, and in those days rarely visited by anybody, the area is tropical, though semidesert, and the waters abound with fish. Less than twelve inches of rainfall tend to fall in just a few days midyear. The Indian Ocean around the Muiron Islands is often very clear, revealing extensive patches of coral and limestone reefs.

I launched the patrol boat's dinghy, fitted the outboard engine, motored to about one hundred yards offshore, and anchored beside a fringe reef of coral. I jumped over the side wearing a wet suit, but the water was so warm that I soon unzipped the front of my jacket to cool down. I was swimming along in about fifteen feet of water and was probably fifty yards or so from the dingy, when I noticed a coral ledge standing about two feet above the rocky bottom. White feelers of spiny lobsters were sticking out from crevices in the rocks. I dived down and could see, right at the back of the ledge, some very large lobsters in the ten- to twelve-pound range. I dragged myself into the cave until I was half a body length under the rock, thrust my hand spear into one of the largest lobsters and started to back out. (Spearing lobster was allowed then, though it is now widely banned).

As I pulled back, my snorkel caught on the roof of the cave. I tipped my head to the side to unhook myself and pushed out farther, pulling the lobster with me. As I came free, I noticed a yellow sea snake, about five feet long with black diamond patterns on its scales, coming straight for my face. Normally, sea snakes, though highly venomous, are not particularly combative—more curious than anything else. Although they're abundant in this region, I'd never had cause to be concerned by them, however I did pay close attention to make sure I didn't upset any. In June and July, during the mating season, it's been reported that sea snakes can become aggressive, and this one seemed to be living up to that reputation.

There was very little I could do. Although I was wearing a wet suit, which wouldn't have been penetrated by a sea snake's short fangs, he was coming straight at my exposed face and neck. At the last second, when he was no more than twelve inches away, I grabbed him behind the head with my right hand, shoving him from my face. He was round and thick with a paddlelike flattened tail. The snake was a powerful animal and started writhing and twisting, wrapping himself around my arm and body. I held on tight; the snake held on tight. I dropped the lobster and spear and started for the surface. The struggling snake impeded my motion, so by the time I burst through the surface I was gasping for air. This animal was not at all happy with me and, if I wasn't supercareful, could kill me in minutes with one bite. I rolled over onto my back and finned my way to the dingy, all the while grasping the snake tightly to make sure that his fangs didn't get near my face or bare chest and hands.

I got to the dingy and hung onto the boat with one hand while gripping the snake with the other. I rested for a minute, then started whacking his head against the side of the aluminum hull—and I kept whacking. Its muscles tightened even more and I had difficulty getting him to flex enough so that I could hit him hard enough against the boat. At the same time, I was cautious about the possibility of a broken fang penetrating my hand. After a few minutes, there wasn't much head shape left, even though the snake was still writhing. I discovered that a snake's muscles can continue moving long after it's officially dead, and I didn't trust the fact that he could no longer harm me. Eventually I decided to lift the snake into the dingy and let it go from my cramping fingers. Now there was a writhing snake in the boat and I was in the ocean beside the boat, pumped with adrenaline. I trod water for about fifteen minutes as I cooled down and let the adrenaline surge subside. It took this long for the snake to stop moving, though it still twitched from time to time.

When I got back to the patrol boat, I skinned and scaled my catch and brewed up a pot of very, very, strong black tea. Tea has a high level of tannin in it and the old-timers in my youth had advised me that if ever I was stuck in the outback and needed to tan leather, I could use tea. I later learned that this animal was a beaked sea snake—the deadliest of more than fifty different species. One bite can deliver eight to ten times the venom necessary to kill a human. The snakeskin became a valued belt in my wardrobe for many years, though I had no clothes that it went with. What would a man wear with a yellow and black diamond belt anyway?

Just to keep us on our toes, nature evolved snakes from lizards about one hundred million years ago. Intact snake fossils are rare, due to the delicate nature of the reptiles' bones, but in 2005 paleontologists in Patagonia found the oldest snake fossil identified to date. Named *Najash rionegrina*, it definitely shows a sacrum, pelvic girdle, and a type of hind legs.

There are twenty-nine hundred known living species of snakes, ranging in size from the four-inch thread snake to the largest captive snake in the world, which died in 2010 at the Columbus Ohio Zoo. Named "Fluffy," he was a reticulated python, twenty-four-feet long and weighing three hundred pounds. My friend the sea snake was a descendant of venomous land snakes that migrated into the oceans around five million years ago. Their adaptations to an aquatic environment include a tail that is paddle shaped to assist with swimming, a much larger lung (snakes only have one lung) than terrestrial snakes so they can stay underwater for long periods, and little flaps or valves in their noses to close off their nostrils when they're submerged.

I didn't realize while I was wrestling with the snake that he was demonstrating a unique feature that is now being studied for application to chemical and gas filters. Identified by researchers in the 1980s, it was described by noted biomechanics research professor Steven Vogel of Duke University. Sea snakes breathe air like we do, yet they can hold their breath for up to two hours, dive more than two hundred and fifty feet deep, and swim for long distances under water. Studies on one sea snake showed that it takes a deep breath and descends to about ninety feet below the surface; swimming down against the buoyancy of its expanded lungful of air. As the water pressure increases with greater depth, the lung's volume decreases, together with buoyancy, so the snake can swim steadily at that depth. To return to the surface smoothly, rather than being pulled up too quickly by the air still held in its lung, the sea snake loses gas, mainly carbon dioxide and nitrogen, through its skin, whose permeability acts like a kind of gill. It can gradually ascend, maintaining neutral buoyancy as it returns to the surface. Reverse engineering the architecture of the sea snake's permeable skin could allow the construction of membranes that permit specific gas molecules to pass through while excluding others. This might be applied to separating carbon dioxide or noxious gases from air, for example.

The U.S. Department of Defense is interested in snakes as well and has funded research on the ability of certain tree snakes to fly. It's not that the tree snake flies as much as glides, since it has neither wings nor stretchable membranes like a flying squirrel. Instead, the normally round serpent is able to flatten his body to create a long, highly maneuverable ribbonlike, wing. This strategy could potentially be applied to all manner of devices seeking improved lift, which is an engineering term for the response to the difference in pressures above and below an object, such as a wing. If the lower pressure is above the object, the object will rise against gravity. In the same way, when you suck on a straw and lower the pressure within it, you can pick up a piece of ice or a slurp of drink. The military is not letting on exactly how they might use flying-snake strategies in their operations, but it may be revealed in future surveillance or battlefield applications.

Consider the pit viper. The name sounds mean and they certainly can be dangerous. The pit in question actually doesn't describe an environment the snake lives in, but the snake's pitted face. These pits on each side of the head are loaded with thousands of sensors—thermoreceptors that detect heat, a lot like the high-tech gadgets revealing the heat signals of hidden humans. In the snake's case, what is even more astounding is that it can see three-dimensionally. A viper can detect the distance and accurately strike a mouse prey in 0.035 of a second—now, that's quick. Scientists at several research centers, including the U.S. Air Force's Wright-Patterson Air Force Base in Ohio, have researched the mechanism by which pit vipers and other infrared-radiation-sensitive animals function. The heat-sensitive pockets on the viper's face contain sandwich-like layers of nerves, capillaries, and even mitochondria that become agitated in the presence of infrared radiation. These work together to signal to the brain the exact dimension and location of the heat source. By better understanding how each of these factors contributes to the animal's sensitivity and precision, the biomimetic application is clear: increased accuracy in three-dimensional infrared cameras. The results would be improved imaging and range-finding for search and rescue, defense, and police departments, just to name a few.

FEATHERING OUR NEST

Galah

"His name is Fang," said the teary lad. "We're moving house and we can't keep him." Fang, the pink and gray galah (a colorful Australian cockatoo known for its incessant and raucous chatter), was brought to my wildlife office in a cardboard box. By the telltale wrinkles around Fang's eyes I guessed he was about ten years into his projected eighty-year life span. "Be careful; he bites really, really hard," said the young teenager, as he handed me the box.

Though galahs were a significant food source cooked into pies, stews, and soups during the Great Depression, it turns out that they are highly intelligent and have the problem-solving abilities of a two-year-old human child. Perhaps their greatest talent, from an Australian point of view, is their ability to tear the lid off a beer bottle with their beaks.

I was used to receiving all manner of wildlife, from pelicans to joeys (infant kangaroos), whether they were injured, abandoned, or had outgrown their welcome with their owners. In twelve years as a Fisheries and Wildlife officer, this was my first galah. I opened the door to the largest bird cage I had, fitted a long leather welder's glove to my hand, and reached into the box. Fang latched onto a finger and bit as hard as I imagine he could, with his curved, sharp beak—evolved to crack the hardest seeds and nuts. Thank you, glove. I brought him home and set him up with cage, toys, and seed on the back porch outside my kitchen, so he could have close contact with me whether I was in or out of the house. He was sullen and silent for the first few weeks, unusual for a species known for its cacophony of screeches and ability to talk. However, galahs mate for life and can also become very bonded to humans, dogs, and cats in their close environment. They suffer anxiety and depression when separated from their habitual companions.

One summer evening a few weeks later, I had guests to dinner—a dignified and formal gathering of charity fund-raisers. We had just sat down to the candlelit table on the poolside patio. In those moments of reserved, slightly awkward silence before wine loosened ties and tongues, Fang suddenly screeched, at full volume, "Fuck off, you bastards! Get the fuck out of here!"

Talk about an icebreaker. Fang didn't stop yelling profanities until I threw a darkening cover over his cage. The next day, I ran an ad in the local paper:

Free pink and gray galah.

Great talker.

Potty mouth.

I was swamped with inquiries. I could only imagine the soup Fang's vocabulary would land him in, in his coming decades.

Nature has a lot to teach us about flight. There is a species of hummingbird that beats its wings six hundred times per minute and flies clear across the Gulf of Mexico on just two grams of fuel. The ultraefficient albatross can fly for many hours at a time and six hundred miles a day. Thanks to specially serrated feathers and extrawide wings, owls are so silent in flight that mice and gophers have no warning of attack.

Every bird is built to nature's streamlined design ratios, so it would make sense to apply them to aircraft. In fact, two of the greatest aircraft designs of all time—the DC3 and the Boeing 747—are built more or less to nature's proportions. But planes don't actually fly. Instead, we add enough horsepower and thrust to stop a far-heavier-than-air equivalent of a brick from falling out of the sky. A Boeing 747 airplane weighs around four hundred tons at takeoff, or 180,000 clay house bricks (roughly ten good-sized brick homes), and burns about five gallons of fuel per mile. San Francisco International Airport, not the largest airport by any means, delivers up to three million gallons of fuel per day to aircraft.

From left to right: Japanese Shinkansen "bullet" train and kingfisher

Moving at very high speeds, Japan's 500 series Shinkansen bullet train produces significant noise, primarily from wind resistance over the train's body and rooftop pantographs (mechanisms for collecting electricity from overhead wires). Eiji Nakatsu, an engineer with West Japan Railway Company, studied the head and beak of birds like the kingfisher, which can dive into water noiselessly and without a splash, in order to streamline the train's nose for a more aerodynamically quiet ride and reduced drag. The wedge-shaped beak starts at a point and expands in diameter at a very precise rate. Eiji also applied the design features of owl wing feathers to quiet the pantographs. Owls have a unique, fringelike adaptation, called a fimbria, on the leading edge of their primary feathers. It appears that this structure changes the sound frequency of their wings passing through the air to a higher spectrum, which is beyond the hearing range of humans—or more important for owls—their prey.

Don't you hate it when you drop your cellular phone, camera, laptop, or other electronic instrument onto the floor? You know the shock can damage or destroy it. The shock of even a relatively small collision can damage not just your iPad but your brain, car bodies, aircraft flight recorders, and space vehicles. A woodpecker, however, can whack his head against a tree up to twenty-two times per second and twelve thousand times per day. He strikes with a massive force of twelve hundred grams of deceleration (scientific term), while humans get concussions at just one hundred grams of deceleration. How does he avoid the monster headaches, detached retinas, and brain injury that humans would suffer from just one such impact? First, he has a supertough yet elastic beak and "spongelike" bones to absorb shock. He also has purpose-built, thick muscles and vibration-suppressing spinal fluid. His eyelids act as "seat belts" to protect his retinas. How do we know all this? Researchers have been studying the woodpecker since the mid-1970s. Yubo Fan, a bioengineer at Beijing's Beihang University; Lorna Gibson of MIT; and researchers Sang-Hee Yoon and Sungmin Park of the University of California, Berkeley have all written scientific papers on the biomechanics and potential of learning from woodpeckers. Sang-Hee Yoon and Sungmin Park have now built woodpecker-inspired biomimetic shock absorbers that can withstand the sixty-thousand-gram impact from a bullet—sixty times the shock resistance of today's advanced aircraft black boxes, which currently can only handle one thousand grams of impact. As you can imagine, helmet makers, such as Riddell, as well as the U.S. Department of Defense are looking at adapting the technology to helmets as well as other applications.

Cormorant

I've always admired cormorants for their incredible underwater swimming ability. They're not the best at aerial flight but can swim faster than the fish comprising their diet. Cormorants are so adept at catching fish that in some parts of Europe and Asia people tether them with a string attached to a ring around their necks, let them capture a fish, and then pull the bird back ashore. The ring prevents the bird from swallowing the fish, and the bird keeper is able to retrieve the fish for his own use. I've heard that the owners highly value their cormorants, form strong bonds with them, and care for them lovingly.

The gannet, or booby bird, is also an excellent catcher of fish. A handsome animal, larger than a seagull, it's mostly white with black and sometimes yellow markings. To hunt, the gannet cruises a hundred feet above the ocean. When he spots a tasty-looking fish, he dives almost vertically. He gathers considerable speed in his fall and at the last moment before impact, he folds his wings back to create the least possible drag. His momentum, assisted by paddling with webbed feet, enables the gannet to plunge thirty feet deep to catch and sometimes even eat fish while underwater. To handle the impact, the booby has no external nostrils to fill with water. Like a modern car, it also has air bags in its face and chest, just under the skin, that cushion the impact. Its beak and skull are highly streamlined and reinforced against shock, while remaining lightweight for flight.

Once our gannet has caught his fish, he usually swims to the surface and flies into the air to devour it. That's when he may lose his lunch to a frigate bird. Large and black, with wings shaped like a World War II Stuka dive-bomber aircraft, the frigate is one of the world's fastest birds—recorded at almost one hundred miles per hour. A frigate doesn't fish for food; he dives on others, such as a gannet, and plunders *its* catch. Although frigates are among the largest flying birds with up to a seven-foot wingspan, their entire skeletons weigh just four ounces—numbers to make NASA and air force scientists drool.

Aircraft designers, carmakers, and public transport engineers are continuously seeking better technology for their air bags and other impact safety features. As Eiji Nakatsu did to improve the noise of trains, analyzing and adapting the bone structures and profiles of birds can be of great value to transportation engineers seeking to build vehicles that are increasingly lightweight and energy efficient, as well as strong and safe.

Tap, tap, tap

"What the . . ." I rolled over in bed. Summer's dawn light showed 5:30 a.m. on my watch. Tap, tap, tap . . . at the door of my government trailer. I was twenty-two years old and just yesterday had taken over my first outback conservation district as a Fisheries and Wildlife officer. This country posting was the bottom of the list in desirability and reserved for the lowest ranking, new sap. The trailer was parked all by itself on a toehold of land at the edge of a desolate, coastal dune system. The nearest house, or fisherman's shack actually, was half a mile away.

Tap, tap, tap, tap. Who would be knocking at this hour of the day? I had no instructions about the district and knew almost nothing about my new circumstances except that I would find and move into the trailer. The previous tenant had left three days earlier. I rolled out of bed and pulled on my shorts. The trailer was typical accommodation for young field officers in remote locations: sixteen feet long, a single bed, small kitchen, a breakfast table with a seat on either side, and a cupboard made up my quarters for the next few months. The surrounding bush and a jerry can of water served as a bathroom. There was no electricity, no air-conditioning, and no heating—but it was dry when it rained and the insect screens kept out most of the ubiquitous flies of the Australian bush.

Tap, tap. I opened the door and was startled to see an emu standing, facing me, completely blocking the entrance, its head about two feet from mine. We stared at each other with me trying to make sense of this picture. An emu, close up, is a comical creature, to say the least. Two big eyes are set in a tiny head separated from its large body by a very long skinny neck, and its body is elevated from the ground by very long skinny legs. Those legs although svelte, are tremendously powerful and help the bird run at speeds up to thirty miles an hour. They can inflict serious damage by kicking.

I was intrigued by my new friend and offered him a slice of bread from my kitchen. He immediately pecked it out of my hand as if he had done so many times before. I repeated the offering several more times until he had satisfied his hunger and then he just continued to stand there staring at me. For whatever reason, I felt compelled to reach out to him. I gently put out my hand to touch his long neck. He didn't move. I started slowly stroking his neck in a downward motion so as not to ruffle his feathers. He took a small step closer to me. I kept stroking for a minute or two and then noticed that Moo, as I christened him, started to quiver. His legs were visibly shaking and the shaking was increasing. A few more strokes of the neck and Moo's legs collapsed under him and down he went to a squatting position. I stopped stroking; he stopped shaking, but just sat there. I watched him for a few minutes more and then closed the door and went back to bed. When I got up again Moo had disappeared. However, the next morning at 5:30 a.m.—tap, tap, tap.

Emu

Incidentally, contrary to a popular myth probably coined about two thousand years ago by Roman, Pliny the Elder, the emu's cousin, the ostrich, does not hide his head in the sand when frightened. With the ability to run at forty-three miles per hour, he is the fastest two-legged animal on earth, and with a kick like a mule, he doesn't need to hide from much. U.S. president Theodore Roosevelt wrote in 1918 that, to survive an ostrich attack, you would do well to lie on the ground. The bird may stand or sit on you, but he can't really kick you in this position. However, an angry several-hundred-pound bird perched on you might cause breathlessness—for several reasons.

Ratites, as emus, ostriches, cassowaries, rheas, and kiwi birds are called, pant like dogs when they're hot. They do this because their lungs and multifolded nasal passages are specially evolved to accelerate airflow through the manipulation of vortices. When air accelerates, its pressure drops and so too does its temperature. These internal passages act as evaporative coolers. And just like camels that can't afford to lose water in their arid desert habitat, the ratite's nasal structure is extremely effective at condensing moisture from their out breath and reabsorbing it for future use. An immediate bio-inspired product that comes to mind is a high-efficiency, particulate precipitator for smokestacks.

Apart from eating, shooting for sport, using feathers in fashion, and carving their bones into flutes, what else can birds teach us? Every aspect of a bird's design, materials, components, and function is a masterpiece of efficiency and an elegant solution for its task. Evolved over the past 150 million years, the 10,000 species of birds that exist today are the result of an incalculable number of experiments, prototype testing, and mutation to perfectly occupy their respective environmental niches.

Scientists at Pennsylvania State University are studying how birds and insects truly fly—which is in large part due to their flexible, adaptable wings. At different speeds and maneuvers, birds, bats, and even flying squirrels adjust their wing profiles to decrease drag and resistance or to increase buoyancy and braking. At Penn State, scientists have mimicked this ability by creating compliant, adaptable spars, or wing frames, covered by a thin skin of scales that can slip over one another. The whole wing is now changeable in flight. When deployed on aircraft, such a wing will result in greater fuel efficiency and much greater maneuverability in flight. Boeing is taking a firm step in this same direction. The company has already patented a morphing winglet and anticipates it will produce fuel savings of up to 5 percent over the duration of a flight. Based on average time in the air for a Boeing 757-300, that could total $164,000 saved per year per plane.

As scientists at NASA Langley Research Center reported in a survey of biomimetic opportunities, although humans can "build eight-hundred-thousand-pound 'birds' that cruise at near the speed of sound (or greater) for thousands of miles nonstop," these machines are greatly inferior to nature's designs in many respects. NASA is reexamining biological flight for inspiration in energy efficiency and noise reduction. Their research confirmed that "natural systems tend to minimize cost for maximum gain, adapt to changing conditions, self-repair, and achieve cooperative behavior for the good of the flight group." Already, biomimetic flight research has revealed many key findings. One report on morphing wings announces, "The potential for broad, significant change in the capabilities of future vehicles remains unbounded." NASA, Boeing, and the U.S. Air Force have now also built an experimental, adaptive wing that changes shape according to conditions.

In the 1970s and 1980s, early biomimetic researchers in NASA Langley's viscous flow division studied the morphology of birds and marine animals—particularly their surface textures. As we know from sharks, the variations in the skin or feathers of animals can set up a layer of tiny whirlpools at the boundary between their bodies and the medium in which they fly or swim. NASA found that adapting these surfaces to aerodynamic and hydrodynamic vehicles produced up to a 6 percent reduction in drag. This approach was also used to create a textured, stick-on plastic film that was applied to the hull of the winning U.S. challenge boat in the high stakes 1987 America's Cup sailboat race series.

What else excites NASA and the U.S. Air Force about birds? Birds are cool flyers—literally. They don't generate heat as they pass through the air—and therefore aren't targets for heat-seeking missiles. Their flexible wings don't crack or fall off. They respond so effectively and independently to changes in air speed and density that they don't get destabilized, spin out of control, and crash. Their wings also totally fold away when resting—which would be a real plus for storage and security, particularly on aircraft carriers.

Birds are also great exploiters of ground effect. It has long been known that aircraft and helicopters, like birds, experience increased lift and reduced drag when about one wingspan above the ground or water. The air between the ground or water and the wing gets compressed, providing an additional upward force to ride on. Seabirds such as albatross, shearwaters, and pelicans use this to great advantage. Special aircraft have been developed using this strategy, particularly by the former Soviet Union, as inexpensive, extremely fuel-efficient transport.

Most birds don't need runways. They simply jump into a flight position and take off, a great advantage if adaptable to commercial and military aircraft. A few, like pelicans and buffleheads, do run a short distance to start and sometimes to land. Ravens and doves can somersault in midair with no loss of flight control. Why, red-tailed black cockatoos even mate on the wing. Imagine if fighter aircraft actually could maneuver as effectively and gracefully as a dove, swallow, or a raptor or as silently and well camouflaged as an owl. It's not surprising that the U.S. Air Force calls one of its highest performing aircraft the Raptor. Although a spectacular technological achievement, it is still no match for biological flight.

Dr. Geoffrey Lilley, a professor emeritus of Aeronautics and Astronautics at the University of Southampton, is an expert on biomimetic, quiet flight. Working with NASA Langley, he confirmed that the raggedy fringe on the trailing, or back edge of an owl's wings creates "a very large noise reduction." These rough extensions break up turbulence in the wake of the bird's movement into little groups of microturbulence, which effectively muffle sound.

The quill feathers on a vulture and many other birds (the ones that stick out of the wing ends) greatly reduce wingtip vortices, with the effect of lowering drag and energy consumption. As early as the 1960s, NASA Langley biologist Clarence Cone published research on wingtip feather designs that showed how they decreased drag in birds by at least 25 percent. A similar approach has now been implemented on many commercial aircraft; you've probably noticed the upturned wingtips on planes you've flown on. Southwest Airlines paid to have them retrofitted—and earned back the investment within two years through fuel savings—even before the major increases in fuel costs of the last few years. Other airlines have since followed suit.

The German company Festo is a leading supplier of industrial control and automation technologies. Founded in 1925, it has made a serious commitment to biomimicry research and has now taken mechanized flight a step ahead. Through a research project focused on minimal material use and extremely lightweight construction, engineers at Festo created an artificial seagull that actually flaps its wings to fly. Combined with an articulated, torsional-drive unit, which is capable of simultaneously twisting and turning, the bird is the closest yet that humans have come to natural flight. It even looks like a large seagull in its appearance and motion. Festo has also demonstrated its insights into both controls and structure by creating a beautifully graceful, flying jellyfish.

Festo Smartbird

No matter which species or phenomena you study, from the lowest, wriggling life-form to the soaring masters of the sky, from the curl of a bird's beak to the intricacies that enable a snake to fly, science is uncovering a vast array of solutions to challenging human problems. These can provide companies with unassailable competitive advantages and the profits that go with them.

Chapter 6

THE BEE'S KNEES

I used to raise maggots. My mother wasn't thrilled that I left them in her fridge, but acquiesced in order to distract me from other, less sociable, teenage activities I might have pursued. I loved herring fishing, and herring love maggots, so . . .

The end of summer sees the annual migration of Australian herring across the Southern Ocean and up the southwest coast of the continent. It coincides with the onshore depositing of large quantities of seaweeds, shed like autumn leaves. These piles of rotting vegetation attract breeding flies, which deposit their eggs in them. This is referred to as being flyblown. The first storms of the season wash the seaweeds and fly larvae back into the water, which in turn feed the herring. The best possible baits for herring, therefore, are maggots.

Breeding maggots is quite an art, necessitating the placement of a rotten fish on an old piece of sheet iron in the backyard—as far from the house as possible to isolate the stench. Once the meat had been flyblown and the maggots had hatched, fed, and grown for several days, I collected them in a glass jar containing a bed of wheat pollard (chicken feed). In order to keep the pupae from developing into flies, it was necessary to place them, within their jar, into the refrigerator. Maggots, very conveniently, become dormant when cold in the same way we slow the growth of bacteria by keeping food in a refrigerator. In this way I was able to store a whole season's worth of juicy, wriggling bait.

Although surgeons, centuries ago, used to think that maggots were produced by festering wounds, we now know that they're just baby flies. They
generate their ick factor because they feed on dead animals, the deader
and the stinkier the better. Maggots are commonly used for fishing bait
in Europe, where handfuls are thrown into the water to attract fish, with
others threaded on hooks to catch them. But maggots have also been
used for thousands of years as a self-propelled form of wound healing,
since they're quite precise about eating only dead and putrefying cells
while leaving healthy tissue alone. Australian aboriginals and Mayans are
known to have used maggots to clean wounds, and through the U.S. Civil
War in the 1860s, military doctors commented on the startlingly positive
effect of maggots that infested battlefield wounds. The beneficial effects of
maggots fell out of favor with doctors following the advent of antibiotics,
and were more or less relegated to history as an example of an unenlight-
ened, somewhat barbaric, medical practice. However, in the early 1990s,
a small group of researchers, including doctors Ronald Sherman of the
University of California at Irvine, Wilhelm Fleischmann of Bietigheim
Hospital, Germany, and Martin Grassberger of the Medical University
of Vienna, began a systematic study of the efficacy of maggot therapy.
Their published results have essentially given the practice a rebirth. Their
studies found that the chemicals in maggots' bodies kill microorganisms, while their mouths not only secrete enzymes that dissolve dead tissue to debride, or clean, the wound but also stimulate the production of
wound-healing chemicals by the patient. Maggots also secrete allantoin,
a chemical widely used in shaving gels, that has a soothing effect on the
skin. With the increase of antibiotic-resistant germs, maggot therapy has
renewed, practical value and is now used in approximately eighteen hundred medical centers in the United States and Europe.

The medicinal use of maggots is another example of biotherapy. However, maggots and their adult stage as blowflies can bio-inspire materials science breakthroughs as well. The blowfly maggot has skin that
hardens by orders of magnitude as it dries—without needing to be heat
cured, unlike most modern, hardened materials. Julian Vincent of the
University of Bath's Department of Mechanical Engineering, one of the

world's leading researchers in biomechanics and biomimicry, has studied this chemical process with an eye toward manufacturing materials in a water-based solution at ambient temperatures. If deployed, it would allow materials to solidify without the use of the costly, unsustainable "heat, beat, and treat" method.

Researchers in the Evolutionary Biomaterials Group of the Max Planck Institute for Metals Research in Stuttgart, Germany, have also determined that blowflies secrete a fluid out of their feet that acts to increase adhesion when the fly contacts a smooth surface. The fluid, which evaporates without any residue, could inspire nontoxic, biodegradable adhesives for applications ranging from bandages to stamps to envelopes to sticky tape. The feet of blowflies have also been found to have sugar sensors—that become many times more sensitive if the fly is hungry. The underlying chemical process has implications for improved glucometers to measure blood sugar in diabetes patients and the readiness of grapes for wine making in viticulture.

SUPER FLIERS

Those pesky flies that freeload at barbecues and picnics all over the world are actually nature's masterpieces of aerodynamic performance. These Ferraris of the sky have much to teach even our most advanced aeronautical engineers, let alone we poor schmucks trying to swat them away from our hot dogs. Researcher Michael Dickinson of Caltech wondered how these little party poopers managed to so easily avoid being swatted. As he told NPR, by filming with super-slow-motion cameras as he tried to swat, he found that "They perform an elegant little ballet with their legs to position their bodies, so that when they jump, they push themselves away from the looming threat." This entire routine happens in less than a split second. The fly's tiny brain can perceive threat and where it's coming from, calculate the best escape route, complete its pirouette, and leap to safety—all in a tenth of a second. Dickinson calls his research subjects, "marvelous machines—arguably the most sophisticated flying devices on the planet." The implications of Michael Dickinson's research could lead to cars, planes, or helicopters that avoid crashes or car seats that respond to an impending crash, regardless of its direction, by jumping away from a direct impact.

One might wonder what a 1940s style, dark-brown fabric sofa was doing sitting by itself under the scant shade of some thinning trees in the middle of nowhere.

"Used to be a fisherman's camp here," said the fisherman driver of the old truck I was riding in. We were approaching the start of an extensive system of precipitous white sand dunes separating us from the coast. Half-way up the Western Australian coastline and rarely visited, pristine, reef-sheltered Lucky Bay was two miles west as the crow flies. Our fisherman friend wanted to check out the bay to determine its suitability as a mooring for his lobster boat in the coming season.

"Let's stop here in the shade and have a bite to eat before crossing the sand," said Neil McLaughlan, the government officer in charge of Fisheries, a true adventurer and my hero who had kindly brought me on this trip. I had just turned seventeen and fortunately, being skinny, didn't take up a lot of room in the truck's narrow bench seat we all shared. It was summer heat and the exposed skin of my shorts-clad legs was stuck with sweat to the plastic seat. My sweat-wet shirt was sticking to the seat back as well. A rest stop sounded great.

"We can use the sofa," I said.

"Think again," answered Neil. "There are beehives in each arm. They may not take too kindly to you sitting on them." Oh boy, I thought. I could collect some wild honey. I had never actually seen a beehive before, let alone have any idea how to harvest honey. But I had seen in a movie that bees could be sedated with smoke. Lack of knowledge had never stopped me from making foolhardy decisions before, so I set a small fire, covered with smoking green leaves, beside each sofa arm. After a few minutes the bees, which had been flying in and out of small holes in the covering fabric, slowed down and looked positively relaxed. I took a fishing knife, cut the corner of the cloth, and peeled it away from the chair. Eureka! There were about a dozen large combs of honey hanging down from the underside of the armrest top. We had an empty six-gallon ice box in the truck. I sliced through about three quarters of the combs and gently lifted them over the smoking fire. I dislodged any remaining, clinging bees with a stick and placed the combs into the icebox. I repeated the procedure on the other end of the couch and then replaced the cloth to restore the hives.

Delighted with my bounty, I loaded the half-filled icebox into the truck and joined the others for lunch. An hour and a half later we were walking alongside the turquoise water of Lucky Bay, discussing its natural beauty, when Neil noticed smoke billowing up from our former lunch site. Obviously, I had not properly extinguished my fires. Near panic ensued. We ran to the truck and drove hell for leather over the sand hills. If fire spread from this area, fanned by the usual afternoon sea breeze of twenty-five knots (which fortunately hadn't arrived yet), it would be a disaster for the several adjoining two-hundred-thousand-acre parcels of grazing land and their resident sheep. There would be no stopping it; we had to extinguish it now. With wet jute sacks, we threw ourselves at the rapidly spreading fire. It took two hours before we had smothered the last glowing embers. Surprisingly, the couch hadn't burned and the bees seemed unfazed by the events of the day. "Seemed" was the operative word, however. I suspect the hive had dispatched one of the workers to take revenge on me. I was leaning against a tree, downing a mug of water, when said bee climbed up the leg of my shorts and stung me where I remember him most.

It is difficult to overstate the importance of bees to the environment and to the well-being of humans. The more than twenty thousand species of bees are related to ants and wasps and are found anywhere in the world where flowers grow. Ranging in size from less than half an inch to the largest leafcutter bee at one and a half inches long, they're a favorite meal for a range of birds and dragonflies. Yet, of all the bee species, only seven qualify as honeybees. Apart from its obvious use as a food and preservative, honey is a natural poultice and antibiotic and has been used for centuries, if not millennia, as a wound dressing. Several companies, including Honey-Med and Comvita, are now applying honey to the modern treatment of wounds and skin ailments, without the risk of developing resistance to antibiotics.

Beeswax has also been used in every culture of the world, since at least Egyptian times, to create everything from the mouthpieces of Australian aboriginal didgeridoos—an extraordinary wind instrument fashioned from a long, hollow tree branch—to the frets on Philippine lutes to the manufacture of longbows and bullets, to lost-wax metal casting to waxing nails and cotton in shipbuilding to grooming a mustache to candles and more. Interestingly, beeswax can be used as a lubricant or, if blended with paraffin, as an agent for traction. So it can help a sticking drawer in a desk slide more easily or wax a surfboard to ensure the rider's feet don't slip. Beeswax was used as currency in the time of ancient Rome and during the Middle Ages, and the Romans used copper styluses to inscribe characters on wax-covered writing tablets. This early version of a whiteboard could later be softened, smoothed, and written on again.

The third century geometer and astronomer, Pappus of Alexandria, was the first to articulate that the reason bees store honey in hexagonal wax cells is so they can fit the most into the least space. Beekeepers, mathematicians, and engineers since have analyzed honeycombs extensively. A honeybee has to ingest about eight pounds of honey to secrete one pound of wax, so building the most efficient wax structure is a priority. The numbers confirm that a hexagonal shape can best fill, or tile, a given surface area while using the smallest amount of material, in this case, energy-intensive wax. A lattice of honeycomb cells is created via repeating angles of 70, 110, and 120 degrees. Each cell also has shared walls, which conserves more wax as well as fully utilizing a three-dimensional space.

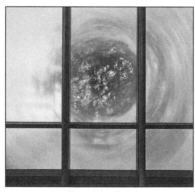

Panelite glass

The New York–based company Panelite has modeled its biomimetic ClearShade insulating glass on the hexagonal structure of honeycombs.

ClearShade controls the amount of sun's heat that's admitted to a building, while still letting in plenty of light. The glass gives an interesting visual effect as well, since its internal structure breaks up the view of the outdoors into an almost pixilated, unfolding image. Installed in the new Jet Blue terminal at John F. Kennedy International Airport as well as a number of university campuses and other buildings, it's able to control heat gains and significantly reduce air-conditioning costs.

Honeycomb structures are also being employed by architects worldwide. A Norwegian firm called Various Architects has applied the shape to a mobile performance space that can hold thirty-five hundred spectators, yet be folded up small enough to fit into thirty sea-cargo containers. The Sinosteel skyscraper in Tianjin, China, designed by MAD Architects, uses honeycomb-shaped windows across its entire facade to passively regulate light and heat. Honeycomb walls also inspired a balcony system that provides both shade and privacy to an award-winning high-density housing complex in Izola, Slovenia.

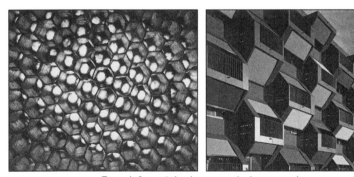

From left to right: honeycomb; honeycomb-inspired apartments in Izola, Slovenia

Ever seen a swarm of bees in flight or coming and going from a hive? They never seem to bump into one another. With a field of vision about 300 degrees wide, bees almost literally have eyes in the back of their heads. This enables them to accurately fly between myriad moving and stationary obstacles. Nissan Motor Company, in its stated quest to halve the number of deaths or serious injuries involving their vehicles, has turned to biomimicry, adapting the design of bees' compound, curved eyes to develop a laser range finder (LRF). When a vehicle is perceived by the LRF to be in acute danger of a collision, the device will mimic the bee's agility and in a split second, take over the car's steering system and turn its wheels to a direction away from the projected impact point. As part of its ongoing Safety Shield R & D program, Nissan took its biomimetic research further and added the abilities of fish to both travel rapidly alongside one another and manage the distance between fish ahead and behind them. In preparation for use in cars, Nissan has created a set of EPORO robot car prototypes that are able to move in unison while communicating to avoid collisions. EPORO stands for "episode 0 robot," since Nissan's ultimate goal is to build cars that create zero collisions as well as zero emissions.

Bees are also key subjects of research on swarm logic, whereby the behavior of a large group of individuals—like a collective of bees, ants, slime molds, or even the layout of a city neighborhood—emerges not because of some central intelligence but due to a cascade of simple signals that trigger actions adding up to complex results. Software is a natural place to apply this strategy, as REGEN Energy of Toronto, Canada, has done to improve the energy efficiency of buildings. They rent or sell building owners wireless devices and software that listen and learn the power cycles of the building's appliances and heating or cooling system. Then the devices coordinate to turn themselves on when needed and off when not needed, in order to reduce overall power use during times when electricity costs are highest. A typical building would have ten to forty controllers that work together as a single hive. Testing suggests that shopping malls, hospitals, hotels, or factories, can save as much as 30 percent on their peak-demand electricity fees.

It's been the common wisdom that fluid dynamics equations "prove" a bumblebee can't fly. But fly they do, with tremendous accuracy and navigational abilities. This impossible ability is the subject of several research programs, including that of Nissan Motors. If humans can duplicate a bee's flying efficiency and maneuverability, aircraft design will take a huge leap forward. Bees also have two-part wings, which zip together for flight and unzip to fold up when landed. If airplanes had wings as maneuverable and foldable as those of bees, then planes would be easier to land and store. This would be particularly useful on aircraft carriers or in airports located in increasingly dense urban areas.

Not only bees' wings are inspiring innovation, their eyes are as well. Wolfgang Stürzl and his colleagues at Bielefeld University in Germany were designing a camera with the widest angle lens possible, which was to be used on a small robotic aircraft. They found their model in a bee's eye, with its stunning 300 degrees of nearly all-round vision. Bees also have a type of "optic odometer," or internal distance measurer, that works by tracking how quickly the image of the landscape moves across their eyes as they fly. Researchers, including Marie Dacke, of Lund University in Sweden, and Mandyam Srinivasan, of the University of Queensland, are analyzing honeybees' navigational capacity with an eye to optimizing GPS systems and tracking devices.

"The bee's knees" was flapper slang from the 1920s to describe something that was considered excellent. The Australian government research group CSIRO has found that fleas' knees are pretty excellent, too. CSIRO has manufactured a "near-perfect" rubber with a 98 percent level of resiliency. Their scientists did this by studying resilin, a protein that makes up the joints of many insects, including bees'. Think of resilin as a type of spring that can absorb force or pressure applied to it and then release that stored energy when the pressure is lifted. Resilin's efficiency is dramatically higher than that of synthetic or natural rubber. It gives bees the ability to flap their wings one thousand times a minute (or five hundred million times during their lifetimes) and allows fleas to store enough kinetic energy to leap one hundred times their body length in a single bound. (The human equivalent would be about 600 feet, giving new meaning to the expression, "the joint's jumping.") It's the most efficient elastic protein known, and its synthesized form could be used to improve everything from the responsiveness of heart valves to the bounciness of running shoes.

WEB OF SAFETY

You've probably heard claims that spider silk has a higher strength-to-weight ratio than the finest steel. In fact, at five times that of steel, it's the strongest fiber known to science. It is also waterproof and stretchable with twice the elasticity of nylon. The ancient Greeks used spiderwebs to cover wounds—not a bad idea, as the spider coats his silk with antiseptic agents. (Just be sure the spider isn't home.)

A research team at Oxford University has made the study of silk, and especially spider's silk, its main focus. Professor Fritz Vollrath leads the Oxford Silk Group, which has generated a number of patents and spun out several companies to commercialize them. Oxford Biomaterials Ltd. was founded in 2001 to develop a silk-spinning process based on the principles used by spiders and other insects to create natural, high-strength fibers. The company plans to use the product of that process: the absorbable, biocompatible fiber, Spidrex, in a wide range of medical applications. Oxford Biomaterials itself has already spun out three new companies to commercialize novel medical sutures; the repair of cartilage, bones, joints; and damaged nerves, including, hopefully, spinal cords. Like most biomimicry companies, the road to implementation of its products hasn't happened overnight, but Oxford Biomaterials and Spidrex represent an excellent example of successful technology transfer from a university setting to the business domain.

It's been calculated that about one hundred million birds die every year from impacts with glass windows—particularly those of large office buildings or skyscrapers. Did you ever wonder how birds avoid spiderwebs? It so happens that spiderwebs reflect ultraviolet light—and unlike us, birds can see ultraviolet light reflections. I guess spiders don't want to bite off more than they can chew—and by creating a seemingly solid object from a bird's-eye view, they also save time by not having to rebuild smashed webs. This clever strategy is not lost on a German company Glaswerke Arnold. Researchers there have developed the award-winning Ornilux bird protection glass. Imbedded in their glass sheet is a chaotic pattern of UV-reflecting strands. These are almost invisible to the human eye but make the window look like an impenetrable barrier to a bird. Bird strikes are down by more than 75 percent in buildings that use their Ornilux product. Glaswerke Arnold, in cooperation with the Max Planck Institute of Germany, has also developed UV-reflecting ink in marker pens and decals that can be applied unobtrusively to existing glass.

Ornilux bird protection glass

I once knew an Australian Fisheries and Wildlife officer named Ross. A wonderful young man, unusually popular, with a wicked sense of humor, Ross was one of three young bachelors posted to a particular country district. The department provided housing for married men but not for singles. However, in this town there was a spare married officer's house, and the chief officer, against his better judgment, was persuaded to let the threesome live in it. Boys will be boys, and after several months of parties, pizza crusts, and a complete lack of housekeeping, the house became infested with cockroaches. Late one night, the three returned home a little worse for wear after a night carousing. Ross was first to enter through the kitchen door and turned on the lights. Large black and brown roaches scrambled in every direction. There must have been dozens. Ross, who was in less than his usual jocular mood, having been turned down by several young women that night, snapped.

"I'm going to get those cockroaches," he said. Out to the garage he went and returned with a five-gallon jerry can of gasoline. This house was not connected to a sewer system. Instead, it had a septic tank under the backyard, capped by a thick concrete lid. To discharge any overaccumulation of volatile methane, a vertical vent pipe went up the side of the house from the effluent pipe to several feet above the roof. Septic tanks provide an ideal breeding environment for roaches. Ross removed a small cap from the concrete lid and poured in the gasoline.

"Watch this," he called out to the others who, by this time, were standing on the back porch, yelling encouragement in the way young bachelors do. Ross struck a match and dropped it into the septic tank.

Kaboom! The lid, or pieces of it, flew into the air followed by an ascending ball of fire. Ross was knocked off his feet. The rain lid on the sixteen-foot-high vent shot into space (and was discovered the next day, more than sixty yards down the street). Back pressure in the effluent pipes blew wastewater back up through the kitchen sink drains and plastered dirty water and scum on the ceiling. It also blew back up the sewer line to the toilet, blasted off the lid, cracked the bowl, and deposited your worst nightmare all over the walls and ceiling. The neighborhood woke up, lights went on, pajama-clad people came into the street. There were many questions about the inferno burning in the Fisheries backyard and the foul stench of a charred cesspool. Interestingly, this treatment seemed to make little impression on the roach population, although married quarters were no longer available to bachelors.

American cockroach

THE ROACH APPROACH

You may have heard that cockroaches might be the only survivors of a nuclear war. Ten million degrees Celsius will vaporize any life-form; however, if a roach was out of heat range, it could survive up to fifteen times the radiation that would extinguish us. There are forty-five hundred named species of roaches on earth and perhaps two to three times that many as yet unnamed. One cockroach likely to cause panic in restaurants and kitchens the world over is the American cockroach, one of that continent's most successful exports. Research has now begun to uncover unique attributes of just a few of these highly evolved species and how they might be mimicked.

We're all aware of the emergence of antibiotic resistant superbugs that pose a growing and serious threat to humanity. While conducting research at Nottingham University, Simon Lee and his research group identified nine different molecules extracted from cockroaches that have proved deadly to bacteria. In particular, they found that cockroach brain tissue wipes out more than 90 percent of antibiotic-resistant *staphylococcus* and *E. coli*, with no detrimental effect on human cells. Why would roaches have this ability? Living in unsanitary conditions means that they're in nearly constant contact with lots of bacteria. They're also coprophages—which means they eat poop. Between where they live and what they eat, it makes sense that they've evolved to cope with microorganisms. Imagine if researchers are able to develop a whole new class of powerful antibiotics that have little or no negative side effects. Investors in those companies could do quite well.

Scientists have been trying for decades to build artificial hands that can wrap around objects, like a coffee cup. Robert D. Howe of Harvard's Biorobotics Laboratory and Aaron M. Dollar at Yale University have successfully developed an artificial hand, fingers and all, based on the dexterity and springing movement of cockroach legs. The hand can already grasp and pick up objects like tennis balls and a coffee cup. Further development holds great promise for amputees to have fully functional hands again. Their innovation led from Robert J. Full's pioneering work at the University of California, Berkeley. More than twenty-five years ago, Bob started studying locomotion, including how cockroaches are able to run so quickly over uneven surfaces. They don't have large-enough brains to be planning their way, so it must be something in their anatomy or their nervous system. Cockroaches have three different types of legs—each with a different function—that work more or less on autopilot so that the cockroach doesn't have to deal with the mental energy requirements of coordinated movement. Bob's research revealed that cockroaches get their speed and agility from particularly springy flexibility in their legs. This allows them to cope with various surfaces without stumbling or tipping over, since an upside-down roach is a very vulnerable roach. His work led to a robot with legs on springs and hinges that scrambles across uneven ground at speeds no robot has achieved before.

Here's a surprise: Termites (called white ants in some countries) are actually members of the cockroach family. Paul Eggleton at the Natural History Museum in London and his colleagues established this fact following genetic analysis of 107 different species of termites, cockroaches, and praying mantises (also a relative). It turns out that termites themselves are interesting from a biomimetic point of view. Their mounds in East Africa and Northwestern Australia can be quite enormous and stand above the sunbaked, semiarid plains by as much as fifteen feet. While the outside environment roasts during the day and dramatically cools at night, the inner rooms and passageways housing millions of the tiny, blind, wood eaters stay at a constant, moderate temperature— without energy expensive air-conditioning. African termites are actually quite sensitive to heat and cold. They also cultivate a particular fungus for food that can only survive at 87 degrees Fahrenheit, give or take a degree. The design of their mounds creates an optimized air exchange within the structure that maintains the delicate temperature balance required. A carefully constructed central chimney allows hot air to rise and escape. That flow also draws cooler air into the lower part of the mound through a series of holes and passageways that circulate the air through chambers underground. They even adjust their vents for cold weather to recirculate heat.

Inspired by termite mound efficiency, Mick Pearce and the ARUP construction company designed the Eastgate Shopping Centre and office complex in Harare, Zimbabwe. The building is passively cooled to match human comfort levels without any air-conditioning system. Ventilation costs are 90 percent less than similar-sized buildings in the same city and saved the complex $3.5 million in energy in just five years. States like California spend more than 25 percent of their electricity on air-conditioning. If passive systems like this can offer up to 90 percent in energy savings, the merits of biomimetic architecture are obvious.

From left to right: Eastgate shopping complex in Harare, Zimbabwe; termite mound

Termites also have an extraordinary ability to eat most of the wood out of a post, house, or bridge without it collapsing until well after the colony has moved out. They know what not to eat—they leave the essential, structural elements. The implications of this skill offer powerful clues for civil and structural engineers—to maximizing stability while minimizing materials and weight in structures ranging from cars to bridges to skyscrapers to aircraft.

There are now over one billion vehicles on the road worldwide, and this number is projected to double within the next twenty years. How on earth will our road systems cope? The average speed of cars in some cities is already less than four miles per hour. Congestion researcher Dirk Helbing and his colleagues at Dresden University of Technology have been taking lessons from an ant colony. They created numerous possible roadways between the nest and a source of sugar. Very quickly, the ants worked out the shortest routes. The whole nest was soon scurrying back and forth collecting food and bringing it home. If this experiment were on humans instead of ants you would expect major congestion, collisions, and road rage. Not so with the ants. No traffic jams formed.

When roadways were at risk of gridlock, some ants diverted traffic away to other routes. This intelligent network is providing researchers clues to better regulating vehicle traffic.

"Hey, boy! Want a cup of tea?" I turned from tightening the mooring lines on the Fisheries patrol vessel. A little corrugated-iron shack was perched precariously on one side of the wharf, overhanging the swampy-brown tidal water, twenty feet below. Standing in the doorway was an old man.

"Are you talking to me?" I asked, wiping the sweat from my forehead, unsure if I'd heard correctly.

"Yes, you. Come and keep an old man company." I looked around the large steel wharf. Apart from the old man, I was alone. My skipper had gone ashore and into the little town four miles away to arrange for fuel. This was a company-owned, iron-mining island—Cockatoo Island—with no private facilities. In any event, it was so remote and lost in the crocodile-infested tropics that no one ever visited unless they were on company business.

"Do you take milk and sugar?"

"Yes, one spoon please." I entered the eight- by ten-foot shed and noticed that it had a swinging sheet of corrugated-iron as a window in its back wall. It was propped open with a forked stick, and hanging out and down to the water below was a heavy-gauge fishing line.

"Caught one hundred thirty pounds of fish out that window yesterday," the old man said. "Here, take a seat. My name's Frank."

Two old, unmatched, wooden stools were in one corner of the hut. A table; a bench with an old army tea urn; and a motley collection of brown-stained, chipped enamel cups lined one wall. A rusted, small bar fridge and a well-worn, floral-patterned armchair, which the old man flopped down into, completed the contents inventory.

"Why are you visiting me?" he asked abruptly.

"Um . . . because you invited me?" I tentatively replied.

"Tell me everything you know," demanded Frank.

I was wondering what I had gotten myself into.

"I don't know much," I ventured. "I'm only eighteen."

"Good, good. That's so good," Frank enthused. "Most young men know everything. Can't learn anything if you know everything. I make tea for the wharf workers—when there are any. Ship only comes in once a week. I'm seventy-eight, been here for ages. They can't get rid of me." He sipped his tea. "Ah . . . that's good. " He scratched his poorly shaven sun-dried cheek. "I see that you're a man with a purpose in life. Do you know what it is yet?"

"I'm just a cadet Fisheries inspector," I responded.

"Want to make a difference, do you?" Frank didn't wait for a response. "Don't let anyone ever tell you that you can't do something if you believe in it. I was a Japanese prisoner of war in Singapore for four years and you know what? I created the world's largest collection of Asian butterflies while I was locked up inside Changi Prison. My parents were taxidermists in India so I knew how to keep butterflies from rotting. Passed them through the fence to the locals and they kept them for me until after the war."

I had never met anyone like Frank. He talked nonstop for the next hour. Frank offered, "If you want, you can come and have dinner with me this evening, and I'll show you some of my collection from this island. They reckon I'm one of the world's leading authorities on butterflies. What do they know?"

That evening my skipper dined with one of the company managers. I excused myself and went off with Frank to his company-supplied, cyclone-proof, corrugated iron house in town. Inside was a splendid display of insects and butterflies, all laid out in perfect order, pinned to display boards through the center of their bodies. Each was labeled. Frank talked at length about the individual attributes of various bright-colored butterflies and moths.

"What's the most interesting thing about these bugs?" asked Frank. He didn't wait for an answer. "You need to see nature as it truly is. You have to look closely to find her secrets. The old aboriginals know about them. People have spent their lives trying to understand what they're about. You can, too." He trailed off as if another thought had overtaken him. "You keep noticing them and one day you might find out something really special. Okay. That's it. It's my bedtime. Good to meet you boy, and stay true to yourself." He ushered me, somewhat bewildered and still unfed, out of his door and quickly shut it. I felt strangely excited. Although I didn't yet understand why, it was an indelible moment that I would remember many times.

LET THERE BE LIGHT

Butterflies have fascinated humans since antiquity and been used for jewelry, art, and decoration for at least three thousand years—as evidenced by Egyptian hieroglyphs. The name *butterfly* seems to have been derived from a pre–eighth-century belief that they stole unattended milk or butter. In fact, their German name, *milchdieb*, means "milk thief." There may be as many as twenty-eight thousand species of these stunningly multicolored, metamorphing animals, with life spans of between one week and a year, depending on the variety. These insects are essential pollinators and now, with the crash in bee populations, are even more critical to all forms of agriculture.

The brilliant colors on the wings of many butterflies and some bird feathers, such as peacocks, are not, in fact, due to any actual pigment in the wing material itself. Instead, color is created through the prismlike, crystalline structure of the surface. The light is split into its various bands of color and reflected to the eye of the observer, in much the same way as the perception of a rainbow.

This effect has been copied by biomimics to produce nonfading, pigment-free paints and electronic display screens. Qualcomm has copied the butterfly effect in its mirasol and IMOD displays. The same layered structures that give a butterfly's wings such vibrancy are used to create an always-on effect without draining energy for backlighting. Because these screens rely on ambient light rather than illumination, colors intensify outdoors, unlike traditional screens that are washed out by daylight. Since the screens don't generate light, they also use about 90 percent less power to operate. If the technology was applied to plasma television screens, for example, which use about 400 watts to run, a Mirasol screen would potentially use only 40 watts. Mirasol technology has already been applied to cellular phones and other user interface devices.

Japanese researchers at Teijin Fibers Limited studied the scales of the South American morpho butterfly, as well as peacock and bluebird feathers. Joint research with Nissan Motors and Tanaka Kinzoku Kogyo began in 1995 and the result was Morphotex, which uses only fiber structure and light reflection to create its color. The textile is made of about sixty polyester and nylon fibers, arranged in alternating layers. Variance in thickness of the strands allows the fabric to look red, green, blue, or violet. Commercial sales started in 2003 and it's said to be "the world's first fiber with a color development system based on a unique structure." Nissan used Morphotex in car upholstery in 2000, though most of the fabric's applications are in clothing. "Artificial leather is another hopeful field for Morphotex," according to a Teijin official, and powder has also been developed for auto paints and printing products. Teijin is a large Japanese corporation with a clear focus on sustainability and reduction of impact on the environment. Its company mission to create high-performing, nontoxic products that can be completely recycled is not just on its website but is reflected in its daily operations.

ChromaFlair pigment

JDS Uniphase Corporation is a specialist in the design and manufacture of products for the optics industry, including optical networks, lasers, and measurement equipment. It created stir-in pigments for paint called ChromaFlair that use the same system to change color depending on the light. The paint system is now used by Dupont, PPG, and many other companies.

African swallowtail butterfly

Another valuable innovation in light also comes from butterflies. Light emitting diodes (LEDs) offer great potential for energy-efficient lighting. If fully adopted, LEDs could reduce the world's lighting energy bill by 80 percent. But the light emitted from standard LEDs is somewhat hard to direct, so it either stays in the LED structure or is sent sideways. Researchers have been developing microtubes that channel the LED light to where it's wanted. African swallowtail butterflies do have pigment in their wings, and in 2005, Peter Vukusic of the University of Exeter found that their brilliant fluorescent color is created in exactly the same way as an ideal LED—and has been for thirty million years. Besides microholes that act as channels to direct the light outward, the butterfly has also evolved a kind of mirror at the base of the channels to reflect any light that gets angled down. As Dr. Vukusic told the BBC, "Unlike diodes, the butterfly's system clearly doesn't have a semiconductor in it and it doesn't produce its own radiative energy. That makes it doubly efficient."

Butterflies also benefit from the minutely rough surface topology of their wings, which ensures that water can't drench or stick. Just as on a lotus leaf, which we'll meet in another chapter, water doesn't soak into the material that makes up a butterfly's wing but rolls up into tiny balls, attaching to dirt particles along the way, and washes the wing clean—a major advantage for a large-winged creature that has no way to groom itself. The world's fleet of aircraft is currently flying thousands of tons of dirt around, at a great energy cost, and use of these nanosurfaces could have major benefit and application to aircraft paints.

Harvard University is developing a somewhat similar approach to prevent ice from forming on aircraft, roads, bridges, power lines, vehicles, and pipes. Instead of butterfly wings, Harvard's Aizenberg Lab was inspired by the surface patterns on mosquito eyes and water strider insects. In these creatures, certain micropatterns limit the adhesion of water. The benefits of biomimicry in this case could mean no more application of expensive and corrosive salt and toxic chemicals to vehicles, roads, and infrastructure—and no more runoff of these deleterious materials onto soil, rivers, and lakes.

DON'T NEEDLE ME

Mosquito is Spanish for "little fly." For most humans throughout history, the only good mosquito is a dead one. However, this tiny but ubiquitous interrupter of sleep, picnics, and meditation is a treasure trove of technologies for biomimics. Evolved over the past thirty million years, these insects have become experts at finding us to feed on. Actually, both sexes mostly drink nectar and act as pollinators for a number of plants along the way. It's only the female that sucks blood—just before she lays her eggs at the end of her life cycle.

Mosquitoes can smell and hone in on the carbon dioxide and lactic acid in our breath and sweat from one hundred feet away. They can see movement, which, from their point of view, means life and therefore a tasty blood supper. In addition to these skills, mosquitoes are heat-seeking missiles. Warm-blooded animals and birds trigger heat-detection sensors in these minuscule raptors.

For most of us, there's something in the sight and feel of a needle going into our bodies that elicits a disproportional physical response. When receiving an injection, around 20 percent of the general population either faint or feel like they're going to. No matter how thin or sharp manufacturers make them, needles still sting. Conventional thinking has been to produce sharper, smoother, and thinner needles. However, this approach still irritates nerves due to the needle's large surface area.

Nature, in her constant quest to ensure survival of a species, determined that mosquitoes would probably be annihilated if they caused similar pain. Instead, most species produce very little sensation from the actual skin piercing. Mosquitoes have evolved a proboscis that is serrated, a little like a bread knife, which touches far fewer nerves. The reduced contact area means much less pain. This point was not lost on researchers at Kansai University in Osaka, Japan. They mimicked the mosquito's proboscis and created a tiny, serrated, silicone-dioxide needle just one tenth of a millimeter in diameter—the width of a human hair. When fully commercialized, this needle could be ideal for pain-free blood monitoring for the world's rapidly expanding number of diabetics, which is already 30 percent of some populations and could be that high in the United States within thirty-five years.

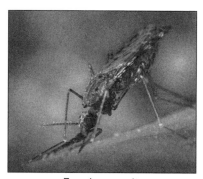

Female mosquito

Disease-carrying mosquitoes are one of the greatest killers of humans annually. Approximately 100 of the world's 2,700 mosquito species are known to carry lethal or debilitating diseases. These include malaria that kills up to one million people each year, 90 percent of whom are children under five years old. More than 3.3 billion people in about one hundred countries are under continuous threat of malaria. The word *malaria* comes from an Italian expression *mal aria*, or "bad air." The disease was described in Chinese medical records four thousand years ago and Sanskrit writings concluded that it was related to insect bites, but the Western world ascribed it to bad air in swamps and wetlands. It wasn't until the late 1890s that Sir Ronald Ross confirmed mosquitoes' role in the transmission of the malaria parasite. That said, dengue fever might actually eclipse malaria as the most significant mosquito-borne disease affecting humans. Yellow fever, Rift Valley fever, and at least a dozen different virus encephalitides round out the top of the list.

Mosquitoes breed in stagnant water that contains a quite precise ratio of dissolved nitrogen and oxygen. Disruption of this ratio can curtail breeding. The challenge is to modify these levels of nitrogen and oxygen in an effective, low-cost manner. Nature achieves this through water movement—either by wind and wave action or by upwelling currents. By accelerating the natural upwelling in water bodies, researchers at PAX Scientific are developing a biomimetic technology for the prevention of water stagnation and the modification of oxygen-nitrogen ratios. PAX biomimetic mixers (we'll discuss these in more detail a little later) circulate large volumes of water with very low energy requirements. As an added bonus, movement of the water surface causes a mosquito pupa, which breathes air through a tiny snorkel, to drown. This technology could significantly reduce the suitability of water bodies for mosquito breeding, is chemical and toxin free, and can run on solar power.

The malaria-spreading *Anopheles* mosquito turns out to be the preferred cuisine of the East African jumping spider. Found in one of the world's most malaria-prone regions around Lake Victoria, this little spider could well inspire future biomimetic malaria control and detection. And how about mosquito birth control? Male mosquitoes were thought to be deaf over certain frequencies, but new research shows that both male and female mosquitoes court each other at very high frequencies. Different types can even be recognized by their whine. This offers biomimics the opportunity to replicate these frequencies in nontoxic, sonic, mosquito traps, or to create sterile mosquitos that successfullly attract females but reduce populations overall due to the associated reduction in breeding.

BEETLEJUICE

There are insects that even offer help with the world's water issues. One bug has worked out how to catch and drink morning fog in one of the driest places on earth. The bodies of Namib Desert darkling beetle species are covered with little bumps, about twice the thickness of a human hair, while the shell itself is covered with a waxy substance that is highly hydrophobic (repels water). This little fellow sticks his butt in the morning air, so that minute water droplets are attracted to the bumps and then run down his back and into his mouth. Designer Kitae Pak has created an award-winning biomimetic device based on the beetle's strategy. Made of stainless steel, the beetle back–shaped dome collects droplets from Namibian fog and runs them into a circular reservoir for later consumption. This could potentially provide enough water per day for survival of Namib Desert peoples.

There is a semiaquatic rove beetle that interacts with water in a very different way from the Namibian desert beetle. As the Ask Nature database describes, when speed is required, it skims across the surface of ponds powered by a unique propulsion system. "Abdominal glands of the rove beetle help it skim quickly across water via secreted chemicals that locally reduce surface tension." The chemicals react so violently with the water that the beetle is shot across the surface at high speed, not quite like a released balloon, but you get the idea. Here is a plethora of biomimicry opportunity. Mimicking these molecules could create a nontoxic treatment that, by reducing surface tension, could drown mosquito pupae by preventing them from breathing through their snorkels. Another application is to reproduce the secretions of a rove beetle and put them on ships or hydroturbines in order to reduce friction under the constriction of surface tension. It's even conceivable that ship propulsion could be developed that didn't need motors or propellers.

WAY OUT INSPIRATION

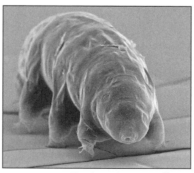

Tardigrade

Is there life in space? One thing we do know, thanks to the European Space Agency, is that there are earthly life-forms that easily survive and even thrive in the extremes of space. Apart from some species of lichens and bacteria that have taken rides on space shuttles, there is at least one animal that has laid eggs 160 miles above the earth's surface. Tardigrades (also known as waterbears, moss piglets, or their affectionate nickname, tardies) can withstand temperatures as low as minus 273° Celsius (which is minus 573° Fahrenheit and nearly absolute zero) and as high as 151° Celsius (303° Fahrenheit—one and a half times egg-cooking temperature) without a shiver or a sweat. They can live in a vacuum, liquid helium, or 90 percent alcohol and are impervious to radiation one thousand times anything that would kill us. Tardies are teetotalers, not needing water for a decade at a time. They can dry out to just 1 percent of their normal water content before their next drink—and then probably binge. Wherever you travel on this earth—from Mount Everest to Antarctica or thirteen thousand feet deep in the ocean—you can find water bears. All you need is a microscope. Discovered by German scientist, Johann Goeze, in 1773, there are more than one thousand species growing to a hulking one tenth of one millimeter long (the width of a hair) and in many colors from red to green. They seem to enjoy romance, with very complicated courtship and mating rituals.

Who knows how many secrets we can uncover from these eight-legged little guys about new materials, surviving in space, and even the essence of life itself, but their ability to survive many years without water has made tardigrades a mentor for biomimetic vaccine-preservative research.

The invention of vaccines is one of the great achievements of human ingenuity. Through their use, smallpox has been eradicated from the world, polio is getting there, and literally billions of people have been spared other debilitating or fatal diseases. One major challenge to their wider use in the poorest, most populous, and greatest suffering regions of the world is the absence of refrigeration needed to preserve live vaccines. Half of all vaccines sent to these areas don't get to patients. *Anhydrobiosis* is the Greek-derived word for "life without water." Some organisms, like tardigrades, survive by replacing their water molecules with specialized, stable trehaloses sugars. This allows them to stay dormant but viable for greatly extended periods. Imagine if vaccines could be kept in a dry state with no need for refrigeration. Many more people in remote communities could be vaccinated.

Another organism that we call a bug—a bacterium—is not to be sneezed at when it comes to biomimicry. These microscopic titans create proteins called enzymes that can digest just about anything, including radioactive material. There are germs that eat gasoline and diesel fuel, or neutralize some of the most powerful plastic explosives. Other bacteria quickly evolve to safely eat foods made with genetically modified organisms (GMO) that were originally designed to kill them. A new species of bacteria that eats iron was recently found on the *Titanic*, two and a half miles under the ocean's surface. Named *Halomonas titanicae*, it was discovered by scientists from the Ontario Science Center, Dalhousie University in Nova Scotia and the University of Seville. Within the chemistry of these bacteria lie secrets that researchers can study, synthesize, and ultimately put to use in everything from rust prevention on oil rigs, pipelines, and ships to the cleanup of toxic waste.

Traditional mining strategies for capturing metal ores are often damaging to the landscape and the fragile ecosystems living there. The extensive use of arsenic in the extraction of gold is just one example. While studying microbial systems at McGill University in Montreal, Dr. Irving DeVoe, Dr. Bruce Holbein, and their research team observed the ways that microbes selectively collect, manage, and transport metals such as iron, copper, and zinc, because each has a particular nutrient value to the microbe. By mimicking their strategies, an automated, modular, decontamination and metal reclamation system was developed that can extract forty-two targeted metals including gold, mercury, cadmium, and copper, as well as radioactive and toxic metals. The system can be used to reclaim metal from mining tailings and industrial sludge and to detoxify and remediate water and sewage.

When it comes to treating sewage, the evocatively named poo-gloo looks like a black, six-foot diameter, plastic igloo. Seven smaller domes lined with biofilms of bacteria, which like digesting wastewater, are nested in each device. Wastewater Compliance Systems, Inc., the makers of the poo-gloo, officially changed the name of these systems to bio-domes in 2010. According to the company, a one-horsepower system can replace a forty- to sixty-horsepower conventional system while exceeding its performance. The greatest differentiator between their devices and competitor products is cost. Adding bio-domes to an aquaculture farm or industrial factory is orders of magnitude less expensive than installing a full water treatment system. In existing municipal sewerage systems, poo-gloos can circulate wastewater to improve the effectiveness of treatment. This is valuable as standards for water quality tighten, while municipal budgets are also tight and funds aren't easily available to overhaul an entire treatment plant.

Whether insects, which make up 54 percent of all life-forms, tardigrades, or bacteria, bugs of all shapes and sizes have survived for millions of years. Their hardy structures, complex chemistry, and ways of moving offer a wealth of opportunity to industries from transportation to medicine to metallurgy.

When I bumped into Dr. Daniel Bedo, a boyhood friend who I hadn't seen for many years, I learned that he had become one of the world's foremost authorities on the chromosomes of blackflies. He had spent his career studying them. I asked him how there could possibly be enough happening in a bunch of microscopic chromosomes to keep him fascinated. His response: "When you enter the world of a single chromosome, it's like you walked through the door to an entirely new and unique universe. The lessons nature has to teach us are infinite."

Chapter 7
....

SPORES AND SEEDS

FUNGUS AMONG US

Bouquet of mushrooms

Y ou may have heard that the DNA of fungus is more closely related to human DNA than to that of plants. Did you know that the world's largest, and probably oldest, living organism is a fungus colony living under the Malheur National Forest in Oregon? It has spread its web, or mycelium, three feet deep, under twenty-three hundred acres of land, and is estimated to be between nineteen hundred to eighty-six hundred years old. Now another type of fungus, the slime mold, and in particular the species *Physarum polyceph-*

alum, is inspiring Japanese scientists to make technological systems such as computer and communications networks more robust. When fungus spread out to locate new food sources, their pathways model ideal, optimized interconnections that can teach us lessons about the design of everything from water pipe infrastructure to the World Wide Web.

Paul Stamets is a colorful, bearded, man of the forest. The first time I met him he was wearing a dapper felt hat—that turned out to be made from mushroom fiber. This deeply knowledgeable, passionate force of nature gets standing ovations for his speeches on mycelium at international forums. His talk on the six ways that mushrooms can save the world has been voted "best TED talk of all time" by the online community that has grown up around the TED (Technology, Entertainment, Design) conference.

What's the difference between a mushroom, mycelium, and fungus? Fungi are not plants, animals, or bacteria. They are a kingdom with an estimated 1.5 million species, including molds and yeasts, only 5 percent of which have been identified. The mushroom is the "fruit" and "seeds" of a fungus, but the mycelium is the essential, rootlike part of the organism that lives out of sight, getting its food by decomposing organic compounds via hairlike strands that are as fine as one hundredth of a millimeter in diameter.

Several key moments in life drove Paul to his career as a mycologist. "I was a rebellious kid in a small Ohio town, when I became fascinated by people's fear of mushrooms," Paul told me in a recent conversation. "Mostly, it seemed to me to be fear of the unknown—so I wanted to know everything about them. Puffball spores were supposed to make you blind if you got them in your eyes, but my twin brother's sight seemed just fine after I pelted him with puffballs."

Psychoactive psilocybin mushrooms were taboo, yet Paul's older brother brought home stories of magic mushrooms that were part of ancient mystical and healing traditions in Colombia. The same types of mushrooms were mentioned in a book on altered states of consciousness that this brother brought home from Yale. When Paul lent that book to a school friend, his friend's father burned it. As Paul explains, "That acted to solidify my feelings. If something invokes such fear and reaction, there must be something important there."

His teenage curiosity about magic mushrooms led him to wonder if there were edible mushrooms in the woods near his home. The delight of finding things to eat in the wild launched him into the study of taxonomy, or the critical art of identifying which mushrooms are which. Paul discovered that, although they're hidden in plain sight in the landscape all around us, there was very little written about fungi and how to recognize them. The difference between look-alikes is major in the mushroom world—they can be delicious or deadly. He went on to study fungi at university and to write field guides; and he is now recognized as one of the world's leading mycologists. In a thirty-five-year career, he has authored six books and many scientific papers on the subject, including two authoritative works on psychedelic mushrooms.

Paul is happy to name some of his favorite mushrooms, reeling off their Latin names and qualities as easily as a parent describes his own children. He breaks them into categories: edible cultivated, edible noncultivated, poisonous, and psychoactive. Paul was looking forward to collecting one of his favorites from the deep forest that afternoon—*Lactarius fragilis*, or candy caps. They taste so much like maple syrup that he makes cookies with them. Another favorite, though not to eat, is *Amanita muscaria*, the classic red-capped mushroom with white spots that's often seen in cartoons and fairy tales. As Paul described, "It's not likely to kill you if you eat one, but you will drool a whole lot, be completely convinced you're dying, lose your sense of time, and fall into a deep, comalike sleep." Paul doesn't advocate trying it, although he has, but such a specialized effect merits research. One of his favorite deadly mushrooms is *Galerina autumnalis*. It grows in wood chips, is a pretty orange and brown, and as Paul says, "It's so poisonous that it's like handling an unexploded stick of dynamite. It's fun to know that something so small can be so deadly—I like that a lot." Paul is not a boring guy.

I asked him why nature would need to make mushrooms so toxic, and he gave a sensible answer. "Just because mushrooms can be highly toxic to humans, that doesn't mean they're harmful to their natural environment. They're highly evolved, so the chemicals involved must be beneficial—perhaps related to a habitat's immunity to disease. They're friends with many other species and are absolutely indispensable in living ecosystems."

When I spoke with Paul late in 2011, he had recently returned from a conference on new initiatives in medicine (TEDMED), where he discovered that it now costs an average of $1.5 billion to develop a new drug and get it through FDA approval. That is major investment and risk, even for a large pharmaceutical company. Add this to the huge increase in liability as more and more existing pharmaceuticals are being associated with birth defects and other harmful side effects. As a result, fewer drugs are coming to market—just as we're seeing a rapidly growing, critical need for new treatments against evolving, drug-resistant pathogens.

"Fungi are not drugs, but they're much more than food." After his decades of research, Paul believes one of their most valuable qualities can be to fortify the human immune system, as well as the environment's—to actually prevent disease before it takes hold. Prevention has always been a mainstay of Eastern medicine's approach to health care. Paul's research has convinced him that a combination of Western and Eastern medicine, enhanced by an understanding of nature's treasure chest of mycelial solutions, can reduce the incidence of disease—not just the symptoms.

"All sorts of mycelium have antiviral and antibacterial properties," Paul explained. "Penicillin is just one. There are others that are even more complex and sophisticated in their abilities. I've been involved with research on one species for novel antiviral effects. It looks like it can suppress many viruses from bird flu to pox. Other mushrooms have been scientifically found to be strongly beneficial as an end-of-life treatment, helping the dying to deeply relax with positive acceptance and a peaceful transition."

Paul envisions that the interface of computer technology and scientific research will allow us to create and use a comprehensive database of mycelial therapies—to sit down at a computer, for example, and learn which mushrooms can help your particular health challenge. The beneficial active ingredients of fungi can be biomimetically synthesized by manufacturers, so they can work with and not displace the existing pharmaceutical industry. As Paul stresses, "Nature has endless, undiscovered remedies and solutions to disease—all based on synergism—cooperation between species."

Fungi are not just the annoying mold blooms that we might see on overripe raspberries. Without fungi we would have no leavening for bread, no wine or beer, and no blue cheese. Mycelium is also the major operator in nature's recycling stable. Paul's research has generated compelling evidence that mycelium not only acts as a decomposer of all litter in forest floors but also that fungi and molds can remediate a range of problems from fetid, contaminated waste; oil spills; heavy metal deposits; and chemical insecticides to radioactive pollution and medical infections.

Paul's research has demonstrated that mycelia are powerful remediators of hydrocarbon (oil) polluted soil. One experiment divided a large volume of heavily contaminated dirt into separate piles. Oyster mushrooms were grown on one of the mounds. While the control mound remained lifeless with no change in contaminant levels, over eight weeks the experimental pile reduced its hydrocarbon count from twenty thousand parts per million (2 percent of the total material) to around two hundred parts per million. Direct application of fungi, or mycomimetic syntheses of the processes involved, offers tremendous economic value to governments and businesses that need to clean up toxic waste sites.

Perhaps one of the greatest tricks of the mycelial world, and an incalculably valuable model for mycomimicry, is its ability to evolve hugely complex and efficient networks. It does this through radial growth. When a mushroom spore sprouts, it grows a short initial section called a germ tube. That tube grows and branches into arms, or radii. Each of its radii grows and branches again. By branching over and over, the mycelium spreads out rapidly and also develops a circular shape as the layers of branching tubes multiply so often that they meet one another. This results in connections between the outer ends of the branches that allow nutrients to move easily to where they are needed, without having to start at the beginning and follow a linear path. The neural pathways of the human brain—and the Internet—follow a very similar construction as mycelium.

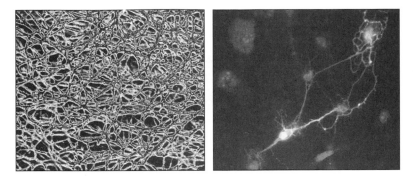

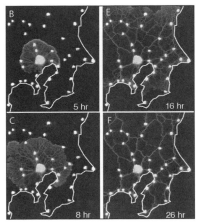

From left to right: mycelium, human brain neurons,
Toyko suburb map and slime mold pathways

Paul is certain that artificial intelligence of the future can be self-educating and mimic the natural networks of fungus. His research suggests that these organisms are aware, responsive, and sentient in ways we don't yet fully understand. In experiments at Hokkaido University in Sapporo, Japan, mycelium was allowed to grow on a map of Tokyo, with tempting oat flakes representing thirty-six nearby cities. To get to the oat flakes, the fungus worked out more efficient pathways than the current Tokyo railway system reaching its suburb cities. When various topographical features were introduced, the similarity to the existing transportation network was even more precise. This strategy could be adapted to improve everything from road planning to more efficient computer communications.

The fact that these organisms might be "smart" is a bit unsettling. But whether or not fungi can think, they are proving their value in medical applications. Results released in early 2011 on a multi-million dollar, seven year study funded by the NIH showed that freeze-dried turkey tail mushroom mycelium supported immune function when taken by women with breast cancer. Their immune response was enhanced by daily use, with none of the patients showing any adverse effects. There is sweet irony to this story. Paul's farm supplied the mushrooms for the NIH grant starting in the mid-2000s. Later, but before the clinical studies were completed, Paul's eighty-four-year-old mother was diagnosed with the same kind of cancer, but far more advanced, as the patients being studied. She took the same mushroom mycelium as part of her treatment program and is now, happily, cancer free.

Paul's business model has mostly been made up of what he calls "guerrilla survival techniques," with initially no employees and for many years just a few, and no outside investors holding leverage. When he started out, he was on a teacher's salary, raising three children, so dreams of a high-end graduate program had to go. Paul founded Fungi Perfecti in 1980. His stamina came from his conviction that he was into something bigger and more important than himself. "Stay on it," he advises, with the motto "refuse to lose."

When asked how his business model evolved, Paul explained, "I owe a weird debt of gratitude to the czar of the scientific supplies department at the college where I was teaching. He was a roadblock, even for a case of petri dishes I needed for my work. So I decided to create my own supply source. I wanted to create a mail-order business, starting with a supply list and directions on how to cultivate mushrooms, but I needed cash to print up the instructions and buy materials. I was teaching classes while writing a book, so I said to my students, 'If you pay for the book before it's finished, it will only cost you six bucks.' I sold five or six hundred over time—and having a good Christian sense of guilt, it made me sure to finish the book. When it was published, it sold for $35, but theirs were already paid for at $6. Along the way I could give people flyers on how to grow mushrooms that advertised my mail-order business and workshops. The book publicized my workshops and mail-order business, and then workshops publicized my mail-order business and my books, and so on. It was hard to start, but it was a synergistic wheel once it got rolling."

Years later, sales of books and products continue to go well. The family business runs a twenty-acre mushroom farm and has a twenty-five-thousand-square-foot laboratory complex and two distribution centers with more than forty employees. Paul also sells licenses to his patents and continues to conduct research, write, and network with other scientists. While he has spoken with many venture capitalists, he hasn't struck a deal with any of them. The terms didn't sit right and he's seen a number of colleagues who lost the value of their life's work to professional investors. "I've seen huge amounts of venture money flushed down the drain on junk," Paul said. And then he paused. "Sometimes I despair a little. How would you feel if you knew there were solutions that could save people from illness and even death, but you aren't getting them to the market? You have the keys to a locked gate—you can see through it, you can open it, but you don't have the financial strength to walk through it.

"I don't like pitching for money," Paul continued. I don't want to seem too eager—but we're losing species and the fabric of life that gives us support for those species. I'm conflicted, because I meet ethical people who have money and connections, but I don't want to be one more person who wants their attention just for what they can give me. I own a hundred percent of everything, but wouldn't it better to own less of something that could get so much more done?"

I asked Paul what advice he would give to someone who wanted to get into biomimicry. Not surprisingly for this philosophical scientist, he answered with a question.

"What do you really want out of life? Most people want to be happy and healthy and have some sense that they have a spiritual life, in whichever faith tradition they practice, and be satisfied that their life's work is in a good business. Is that what you want? If your answer is yes, then you must follow your heart. Let it guide you in the ways you can contribute. Follow your heart and don't follow money for money's sake. If you do this, you'll be rewarded by the gratitude of others who respect your philosophy of life and carry your mission forward, voting for you by buying your products."

And what about someone who is already in the trenches? What advice would Paul give them? "Question all assumptions—because the biggest mistakes are made at the beginning of the decision tree. I should not have been the one who discovered that the mycelium of insect-parasitizing fungi *attracts*, whereas it was well known that the spores of the same fungus *repel*. The assumption of my predecessors was that this wasn't possible. They were wrong, but it took my risking being wrong to discover it. This is important—you need to fail, frequently and often, because the price of failure is the cost of tuition you've paid to learn a new lesson—you have gained new information. The only way is to try. The more you try, the more likely you will have an effect. So dare to fail, because with only one or two successes, you pay for all the prior failures. And the challenges and hardships sprinkled with a little pain keep you alive and fit. They ensure that your satisfaction with life will increase manyfold."

CLEANUP ON AISLE EARTH

While Paul Stamets is a leading researcher on the potential for fungi to solve human problems, other mycologists and biologists are also looking to plants and fungi for help on everything from mining to waste treatment.

The world currently produces about 2,500 metric tons of gold each year. Most gold ores contain only tiny amounts of the precious metal—as little as one tenth of a gram per ton of rock (a ton is one million grams). Ten grams per ton is described as a rich deposit, and just one gram (the weight of a small paper clip) per ton is the average harvest. Every stage of the gold extraction process produces significant hazards. Ore milling produces huge amounts of tailings, or waste rock material, including poisonous heavy metals, like mercury and arsenic, and acid-generating waste that contaminates ground water for decades. A common practice (in 90 percent of gold extraction) is to use highly poisonous cyanide to leach gold from the crushed rock. These materials are then left in open dams and ponds, forming a major threat to neighboring communities of people and wildlife including catastrophic spills and mass poisonings. Cyanide use is banned in a number of countries but still used at most gold-mining sites.

A far cleaner alternative to traditional cyanide leaching is bioleaching, where living organisms are used to extract chemicals. Enzymes produced by iron-activated bacteria have been shown to be effective bioleaching agents. For example, Chile has developed bacterial bioleaching of copper-sulfide ores—and processes more than eighty-five thousand tons per day by this method. Bioleaching can also be accomplished through the use of certain species of fungi that grow on metal substrates. Researchers toured sites where heavy-metal toxins occur naturally in high concentrations. They looked for fungi that thrived on these sites, grew large quickly, and proved to hyperaccumulate the metals. These organisms have been found to generate acids that can even dissolve electronic scrap or catalytic converters with up to 95 percent efficacy. An alternative to bioleaching for waste management is phytoremediation, which uses specific species of fast-growing plants that feed on and or absorb the contaminants in water, soil, and rubble. Leon Kochian of the U.S. Department of Agriculture is a specialist in this emerging field. As he explains, "Certain plant species—known as metal hyperaccumulators—have the ability to extract elements from the soil and concentrate them in the easily harvested plant stems, shoots, and leaves. *Thlaspi caerulescens*, for example, commonly known as alpine pennycress, is a small, weedy member of the broccoli and cabbage family. It thrives on soils having high levels of zinc and cadmium." Although not yet a cure-all for every site, where phytoremediation is applicable it is sustainable and very cost-effective. An additional benefit is that mature plants, laden with the contaminants, can be burned or otherwise processed to retrieve the metals for disposal or recycling. Like biotherapy, these practices directly apply living organisms to solve a problem, but by studying their underlying processes, biomimics could adapt and scale these solutions for even greater impact.

BOBBING FOR COCONUTS

Each year, just fifteen cargo ships spew out as much acid-rain-creating sulfur dioxide and pollution as the world's 760 million cars combined. The global fleet of large ships numbers over 50,000 and uses 370 million tons per year (at about $700 per ton) of the nastiest, filthiest, highest-emission bunker fuel. It smells and looks like liquid bitumen. As mentioned earlier, when seaweed and barnacles accumulate on a ship's hull, they can cause so much drag that they increase the amount of fuel burned on a voyage by as much as 40 percent. Even today's best, very expensive antifouling paints are effective for a relatively short time and degrade quickly. Ships are taken out of the water, at great cost, every two or three years for fresh paint (most pleasure boats—every year). Effective antifouling paint makes a difference to both the bottom line of running costs as well as atmospheric emissions. Unfortunately, by their nature of being toxic to unwanted seaweed and barnacles, conventional antifouling paints are also highly toxic to the marine environment in general. As we saw in an earlier chapter, sharkskin is inspiring a different way of handling antifouling and so, it turns out, are coconuts.

Many of the world's plants, particularly in the tropics, migrate by floating their seeds on ocean currents. Coconuts are a prime example. You can see them washing up on beaches and germinating in the sand throughout the equatorial Pacific, Atlantic, and Indian Ocean islands. It turns out that many of these seafaring seeds don't suffer from the fouling that you see on ship bottoms. In other words, weeds and barnacles don't take a ride on them. This would be a good survival strategy, since accumulation of heavy growth could cause the seed to sink and never propagate. Researchers at the Biomimetics Innovation Center in Germany have discovered that a number of palm seeds' surfaces have microsized, hairlike structures that constantly move, thereby giving no place for seaweed or barnacles to settle. A new antifouling surface has been developed based on the seeds' strategy. The global consumption of traditional, copper-laden, antifouling paint is around eighty thousand tons per year, costing upward of $200 per gallon—a multi-billion dollar market with opportunities and good margins for alternative products.

If you, like many, leave your wet toothbrush in a glass beside your sink, you might have noticed a very thin film of something building up in the bottom of the glass after a few days. That's not just toothpaste that didn't get rinsed off the brush but biofilm. Biofilms are clumps of microorganisms that stick to one another and to all manner of surfaces, causing 80 percent of all human infections.

Many people think of salt water as a good cleansing agent for wounds and infections, but sea and ocean water is teeming with biofilms of bacteria that coat most fish, reef, and coral. Severe and even fatal infections can be the result of an underwater scratch or nick. When I was a nineteen-year-old cadet Fisheries and Wildlife officer, I dropped a lobster on my bare foot. From a tiny prick of my skin, I ended up in hospital for several days with doctors wondering if I would lose my foot due to biofilm-caused infection. Fortunately, I got to keep the foot, but the experience left me highly conscious of biofilm hazard.

BioSignal coating on underwater pylon

Scientists on a marine expedition off the Australian coast noticed that the surface of *Delisea pulchra*, a red feathery seaweed, was completely free of biofilms even though it was living in a soup of bacteria. They discovered that the plants produced chemical compounds—halogenated furanones—that interfered with the way bacteria communicate and assemble into dense colonies. The Australian company BioSignal Ltd. was established in 1999 to commercialize the antibacterial features of those seaweed furanones. An obvious potential application was that ship hulls could be coated with synthesized furanones, instead of environmentally damaging, short-lived antifouling paints. Another area of interest was the interior coating of large oil pipelines to prevent the development of corrosion-causing biofilms. This is a classic example of a bio-mimetic platform technology, one that doesn't fix just a single problem but could make a wide range of existing processes or devices more effective. An outgrowth of the technology transfer program at the University of Sydney, BioSignal was able to utilize university research facilities in its early days and pursued a range of applications including toothpaste to eliminate bacteria-laden plaque biofilms and alternatives to pharmaceutical antibiotics. According to the company, its products have been "used as drug candidates for the treatment of lung infections, as well as in various applications, such as water treatment, oral care products, cleaners, deodorants, catheters, and contact lenses and contact lens solutions." In 2010, BioSignal was purchased by RGM Entertainment Pte. Ltd.

CLEAN AND GREEN

What would life be without flowers? So many of us enjoy growing, arranging, giving, or receiving their beauty. We use them in celebrations of life, marriage, birth, and death. We decorate our churches and bury our dead with them. We don't often stop to reflect that they are the very sex of plants, the only sex organs most societies allow people to handle in public. In common with the sexual parts of all other life-forms, their design and functionality are based on the proportions of nature's spirals. Sap flows in spiraling veins. Leaves are spaced radially around their stems in spirals—the optimum arrangement to share available sunlight. Photosynthesis occurs in spirals and most often, the number of petals and seeds are Fibonacci numbers—nature's ratio of spiral growth. Now an Arizona inventor, Dr. Joseph Hui, has developed a solar panel array in the shape of a lotus flower that even opens and closes.

The lotus is one of the most commercially successful sources of inspiration for biomimetic products. Apart from their intoxicating, heavenly fragrance, lotus plants are a symbol of purity in some major religions. More than two thousand years ago, for example, the Bhagavad Gita, once of India's ancient sacred scriptures, referred to lotus leaves as self-cleaning, but it wasn't until the late 1960s that engineers with access to high-powered microscopes began to understand the mechanism underlying the lotus' dirt-free surface. German scientist Dr. Wilhelm Barthlott continued this research, finding microstructures on the surface of a lotus leaf that cause water droplets to bead up and roll away particles of mud or dirt. Like many biomimics, this insight came quickly, while its commercialization took many years more. The "Lotus Effect"—short for the superhydrophobic (water-repelling) quality of the lotus leaf's micro- to nanostructured surface—has become the subject of more than one hundred related patents and is one of the premier examples of successfully commercialized biomimicry.

From left to right: lotus flower, nanostructures on lotus leaf

Water has adhesion properties. It likes to stick to itself and to other objects. It also likes to live in droplets and spreads out to wet something only if enough of its surface comes in contact with another surface. A superhydrophobic effect is when microsized structures lift water above the surface of leaves, like lotus or kale, preventing it from dispersing and saturating.

More than three hundred thousand buildings in Europe have already been painted with Lotusan, a self-cleaning paint manufactured by Sto AG of Germany and its North American subsidiary, Sto Corp. By reducing water's and microorganisms' tendency to stick in the microscopic cracks of building surfaces, Lotusan paint not only keeps buildings cleaner but reduces the buildup of mold and algae, cutting maintenance costs and increasing the life of the paint job. The effect is also being applied to glass-encased, optical toll-bridge sensors on German highways to reduce the buildup of dirt. Numerous other applications are under development, including the prevention of rain, ice, and snow accumulation on microwave antennas. Just how effective is the Lotus Effect? A spoon with a Lotus Effect surface can be dipped into honey and come out totally clean—without a hint of stickiness. As just one example, perhaps we could reduce the need for high water use and environmentally damaging detergents in dishwashers by keeping our plates and utensils cleaner to begin with.

LEAF POWER

Except for some organisms found in the deepest oceans or the darkest caves, sunlight, collected through leaves, is the primary source of energy for all life on earth. Plants are eaten by herbivores, who get eaten by carnivores, who die and get digested by decomposers like fungus and bacteria, which then become soil for the next generation of solar-energized plants—and the cycle goes on. Photosynthesis in plants and trees captures around ten times more energy from sunshine than all the energy used by humans. Almost every leaf of every plant is a solar panel of sorts, so their structures and chemical processes are an obvious place to learn about solar energy generation. This is a strong area of biomimetic research with a number of innovations approaching the market.

In photosynthesis, leaves use solar energy to split water into hydrogen and oxygen. While burning fossil fuels releases carbon dioxide and other greenhouse gases, burning hydrogen to create electricity creates just water as a by-product. That makes the efficient production of hydrogen a very attractive and valuable research focus. The U.S. government sees such merit in artificial photosynthesis that in 2010 the Department of Energy funded the establishment of the Joint Center for Artificial Photosynthesis as one of its Energy Innovation Hubs. Over the next five years, the center is budgeted to receive $122 million in funding—a testimony to its economic potential.

Researchers at Shanghai Jiao Tong University in China, Saga University in Japan, and the University of California, Davis, proposed creating an artificial inorganic leaf modeled on the real thing. They took a leaf of *Anemone vitifolia*, a plant native to China, and injected its veins with titanium dioxide—a well-known industrial photocatalyst. By taking on the precise branching shape and structure of the leaf's veins, the titanium dioxide produced much higher light-harvesting ability than if it was used in a traditional configuration. The researchers found an astounding 800 percent increase in hydrogen production as well. The total performance was 300 percent more active than the world's best commercial photocatalysts. When they added platinum nanoparticles to the mix, it increased activity by a further 1,000 percent.

Another photosynthesis technology, originally developed by Michael Grätzel of École Polytechnique Fédérale de Lausanne (EPFL), Switzerland, is named a kind of "light sponge." Dyesol, a company based in Australia, has applied Grätzel's innovation to overcome the great expense and weight of using silicone, currently the basis of all commercial solar panels in the world today. The use of silicone is a major inhibitor to the world's large-scale conversion from carbon-based power to solar. Instead, Dyesol mimics photosynthesis via a nanotech titanium product, which greatly increases the surface area that is touched by sunlight. Test results indicate that it produces electrical current many times stronger than that found in natural photosynthesis and can harvest solar energy over more hours in the day and at lower light conditions than competing technologies. As the company describes, "The biomimetic Dye Solar Cell by Dyesol Limited (OTCQX: DYSOY) generates electricity well even in low light conditions such as on cloudy and foggy days and is less sensitive to the angle of incidence of light, so it can be installed vertically on the side of a building to supply energy at the point of use.

Leaf-inspired dye solar panels

A team in the chemistry department at the Massachusetts Institute of Technology has gone a level deeper into plant life to study how plants use photosynthesis to transform the energy of sunlight into food. Solar energy generators provide less than 1 percent of the world's electricity, in large part because dark nights and short or stormy days mean that solar panels only work part-time. Storage of the sun's power currently requires batteries, or liquid salts, which are chemically intense and expensive. According to an article in the Christian Science Monitor, Daniel Nocera and his research team "have developed a catalyst made from cobalt and phosphate, that can split water into oxygen and hydrogen gas." It, in effect, can use electricity from solar collectors to make water into fuel. Their goal is to combine it with fuel cells as part of a solar panel system, which could then cheaply store energy for later use.

"You've made your house into a fuel station," Nocera told *Wired* magazine; "I've gotten rid of all the goddamn grids."

Leaves are also teaching scientists about more effective capture of wind energy. Wind energy offers great promise, but current turbines are most effective when they have very long blades (even a football field long). These massive structures are expensive, hard to build, and too often difficult to position near cities. Those same blades sweep past a turbine tower with a distinctive thwacking sound, so bothersome that it discourages people from having wind turbines in their neighborhoods. The U.S. Fish and Wildlife Service also estimates that hundreds of thousands of birds and bats are killed each year by the rotating blades of conventional wind turbines. Instead, inspired by the way leaves on trees and bushes shake when wind passes through them, engineers at Cornell University have created vibro-wind. Their device harnesses wind energy through the motion of a panel of twenty-five foam blocks that vibrate in even a gentle breeze. Although real leaves don't generate electrical energy, they capture kinetic energy. Similarly, the motion of vibro-wind's "leaves" captures kinetic energy, which is used to excite piezoelectric cells that then emit electricity. A panel of vibro-wind leaves offers great potential for broadly distributed, low-noise, low-cost energy generation.

SEED CAPITAL

Aerospace engineers have known for a long time that the most efficient propeller actually has just one blade. This is because in multiple blade configurations, each blade leaves a churn of turbulent air or water in its wake. The next blade sweeps into this turbulence instead of into undisturbed gas or liquid. The result is lost efficiency. Nature worked this out millions of years ago when it evolved the single-winged samara seed, but engineers have had a hard time designing a balanced single-blade propeller. By studying the samara, however, Evan Ulrich at the University of Maryland developed a mini, single-bladed, helicopter. The wing is so efficient that it can stay in the air much longer than any other design for a given amount of fuel. In fact, if flown to take advantage of thermal updrafts, the device could stay aloft indefinitely. An added benefit—if the power fails, the helicopter auto-rotates safely to the ground, just like the seed. The maple samara has inspired not only helicopters but also efficient ceiling fans by Edward Bae Designs and low cost, light, large-scale fan blades by PAX.

From left to right: maple seed samara; Evan
Ulrich's single-bladed helicopter

Another plant, from one of the hotter places on earth, is inspiring biomedical breakthroughs. The Centers for Disease Control report that more than 2.5 million children under five years of age die each year from vaccine-preventable diseases. Bruce Roser, founder of Biostability Ltd. in Cambridge, U.K., is applying a plant trick to create refrigeration-free vaccines. Like the tardigrade, which stays alive through severe drought conditions by preserving itself with specialized sugar molecules, Bruce and his colleagues have implemented the heat-stable, drought-resistant sugars found in the African *Myrotham-*

nus flabellifolia plant. Successful adaptation to make vaccines shelf stable without refrigeration can save millions of lives as well as millions of dollars annually.

"It will create jobs and bring new prosperity to the region," claimed Western Australia's premier in the 1970s, as he agreed to let mining giant Alcoa strip pristine stands of slow-growing hardwoods and create the world's largest bauxite mine. "We can replant the forests after the ore has been mined," he promised. He did not anticipate that the denuded land would no longer support native species and that the non-native trees that would later be planted were so shallow rooted that they would blow over in a storm. He also did not anticipate that using mining equipment in the forest would greatly accelerate the introduction of dieback fungus, the cause of up to 90 percent of the world's largest hardwood trees being devastated. His announcement, instead, was cause for celebration. Property values soared. Local merchants congratulated one another.

I was appalled. Just a few years earlier, the same political party had released millions of acres of one of the world's most biodiverse areas—wild mulga forestlands—for clearing and converting into grazing and wheat farms. "A million acres a year cleared," was bragged about. Films and advertisements were made and shown around the world—attracting farmers and investments. Now, more than 40 percent of that land has degraded into salt lakes and marshes—entire ecosystems ruined forever, an environmental disaster of huge proportions. And now our leaders were attacking the pristine jarrah forests. To add insult to injury, Alcoa was simply bulldozing the forest, pushing the vegetation into huge piles, and setting fire to them. Thousands of beautiful, healthy, productive trees, all being reduced to ashes and carbon dioxide. The entire habitat for hundreds of species of animals, birds, reptiles, and insects decimated. I couldn't bear it. I loved that forest. To see such waste and destruction was against everything I believed in. As a Fisheries and Wildlife officer, my whole career and every purpose in life was to be in, revere, and protect wilderness areas and their wildlife occupants. I'd spent thousands of hours in these forests and their life breathed life into me. I decided then and there that I had to find a new career—not as a powerless government employee but one where I might be able to make a real and lasting difference.

In the short term, appalled by the waste, I planned to go into the forest and scrounge whatever trees I could from the burn piles. There must be many uses for them. However, the piles were too large and tangled to pull logs from. At that time, members of the public (mainly farmers) could buy limited permits to cut trees into fence posts and the like from specified parcels of forest—although not the largest trees, which went to the mill. I applied to the Forest Department and was duly granted such a license for swatches of the Alcoa land—as long as I was done before bulldozer clearing and burning began. The fee was 40 cents per tree. The state did not place a high value on them.

I owned an old six-ton truck and a tractor, so I took some accumulated vacation time and headed into the forest. The land had been set fire to in the fall, in order to clear undergrowth and make the bulldozers' job easier. Australian forests are used to being burned. They're described as having a fire ecology. The trees, themselves, are quite fireproof, and as long as they are left alone, rapidly spring back to life. For millennia, Aboriginals set fires to encourage new growth—many seeds will only germinate after a fire—in order to attract game animals in to feed on the tender new shoots. This was now winter and a carpet of green covered the forest floor. New life was burgeoning everywhere, unaware of the pending annihilation.

I spent five weeks alone working the forest, chain-sawing twenty-five trees each day. Work lasted from first light till dark every day, seven days a week. Before cutting a tree, I would stop a moment and appreciate its beauty. On very windy days I stayed home; much too dangerous to fell trees when it is windy. I found that out the hard way. As many folks know, to safely bring down a tree, you need to determine if it has a lean or weight bias in any particular direction. Maybe there are more branches on one side than the other. You then make cuts in such a way as to help the tree fall in that direction. This one particular tree—larger than most, was pretty well balanced, but had a slight lean. I cut a wedge shape in the trunk on the side that I wanted the tree to fall. A strong wind was blowing. The tree I was cutting and the surrounding trees were swaying. I stopped and studied the situation for a while—a little uneasy. The professional timber fellers had plenty of tragic stories about accidents. I mapped out my escape route for when the tree started its fall. I was confident in my ability to cut right.

The engine roared. The chain teeth sunk into red wood. Deeper, deeper—suddenly, a strong gust of wind came, the blade jammed, the chain stopped. I had lost control of the fall. The only way the saw can jam is if the tree is toppling the wrong way. The remaining wood cracked loudly. I looked up. The eighty-foot tree was swaying—not falling steadily—and my escape route was no longer reliable. The old-timers say that if you look up and can't see which direction the tree is falling, it's falling toward you. Another loud crack and the trunk split in two for about fifteen feet up its length from my cut. The whole tree pivoted at the top of this split, lurched sideways over my head, and fell to the ground with a thundering crack.

I was standing, shaking, between the stump and the log. Not at all where a survivor might expect to find himself. Almost as bad as this was that the top of the tree had landed on my truck, and its hood was squashed onto the engine. That was it—no more cutting in bad weather. I removed the branches from the truck, unbolted the damaged hood, and left for the day.

I was pleased to use some of the logs when building my new house. Ironically, the rest I provided to the local county to serve as railings around protected parklands.

STICKING UP FOR INNOVATION

More than 50 percent of the world's tropical trees have been cut down in the past 50 years. In the last 200 years 90 percent of Australian and U.S. indigenous forests, together with untold species of flora and fauna, have disappeared. Europe achieved much the same over the last 300 years, and skinning the earth is well under way in the Amazon and Southeast Asian rain forests. Imagine if we permanently destroyed 90 percent of the knowledge in our museums, libraries, universities, and laboratories globally; it would be a catastrophe. What vast repository of information have we already lost from nature that we scarcely paid attention to? With continued extinctions, what do we lose every day of every year? Although there are only approximately 250,000 to 400,000 species of flowering plants and trees remaining on earth and under the sea, they still hold an enormous library of nature's technologies, if we look.

Perhaps the simplest of all lessons to learn from plants is the pattern of their sap flow. It can teach us about the path of least resistance and energy consumption and how to pump water hundreds of feet up (in the case of the tallest trees) without any muscles or effort on the plant's part. The branching geometries of tree limbs and leaf veins; the trachea, nervous system, and veins of animals; and the tributaries of rivers all share the same strategy for optimum efficiency. Researchers have confirmed that if they joined pipes in a more natural branching pattern, instead of in the usual right-angle joints humans have used for five thousand years, they saved 12 percent of the energy normally required to pump water through them. When you consider that pumps currently use almost 30 percent of the world's electrical energy each year, a 12 percent reduction is truly significant. It equates to savings of billions of dollars in energy and millions of tons of carbon dioxide emissions annually.

Load-bearing or protective structures—such as trees, bones, shells, horns, and eggs—are optimized by nature to give the greatest strength while minimizing the amount of energy and materials used to create them. As an article in the online Ask Nature database explains, trees "add wood to points of greatest mechanical load by arranging their fibers in the direction of the flow force." By adopting these techniques, engineers have been able to create software programs that grow stronger structures, more lightweight without sacrificing safety. "For example, car parts and entire cars designed with these principles have resulted in new vehicle designs that are at least as crash safe as conventional cars but up to 30 percent lighter," with resultant increase in energy efficiency.

Grass is also structured not to stress. As biomechanics researcher and author Steven Vogel of Duke University describes, "If a grass leaf is split or notched, it does tear more easily but only in proportion to its reduced cross section; there's just no sign of any significant stress concentration. Do your worst to a grass leaf; it just doesn't go along with attempts to tear it crosswise." If we could mimic the stress-resistant properties of just a single blade of grass, we could build much stronger, lighter, cheaper fabrics for use in applications like the spectacular roofs of the Denver International Airport, in Colorado, and various sports stadiums. Truck tarpaulins, tents, and even the Navy SEALs' inflatable boats could benefit enormously.

Architects are early leaders in the biomimetic adaptations of plants and trees. For example, there are no square trees—for good reason. Wind increases in force the higher you go above the ground. As wind pushes on one side of a typical tall building, it creates a type of suction on the opposite side. Skyscrapers generally deal with this challenge by adding extra support structures, but renowned architect Santiago Calatrava showed the beauty and functionality of copying nature with his design for the Chicago Spire. Unfortunately, the global financial crisis and developer issues placed this project on hold in 2010, not long after construction started. At two thousand feet, it would have been one of the tallest buildings in the world. His use of curving forms, which he drew from the image of the spiraling curl of smoke rising from a Native American campfire or the curve of a seashell, narwhale tusk, or cactus, would have greatly reduced the wind forces that act on square or rectangular buildings. Calatrava's nearly columnlike spire allowed wind to flow around the structure while providing greater structural strength—with reduced materials.

From left to right: Santiago Calatrava's Chicago Spire; narwhals "tusking"

Studies on saguaro cactus by the Center for Turbulence Research at Stanford University confirm that these tall, shallow-rooted plants are exceptional at relieving pressure from powerful desert winds. Studies on seaweeds find the same properties in water. It stands to reason that future industrial applications requiring high strength-to-weight ratios and reduced materials, weight, and drag—such as smokestacks, bridge piers, and aircraft spars—would copy natural designs. Overall weight of pilings can be significantly lowered, further enhancing strength and reducing costs. Building with nature's proportions can also create beautiful designs. The Catalan architect Antoni Gaudi was passionate about organic forms and designed the spectacular La Sagrada Familia cathedral in Barcelona according to the spiraling forms found in plants and trees.

When architect Layla Shaikley contemplated the challenge of housing astronauts on Mars, she turned to an unusual source of bio-inspiration. The durian fruit is highly prized among many for its texture and taste. However, I have seen it banned from Asian hotels because of its phenomenal, overpowering stench. A popular feature at Southeast Asian weddings and celebrations, the durian is mildly psychotropic. Unfortunately, consuming it with alcohol can be fatal—a real party spoiler.

However, it does have a structurally unique, semirigid skin design, which inspired Shaikley in the creation of an inflatable building that protects against harsh elements while providing nonreinforced rigidity. As she says, "In the durian fruit, seeds serve on the inside as individual units, yet function as a whole to hold adjacent seeds in place. Likewise, the interior of my inflatable superstructure houses a series of individual and pressurized volumes that provide space at an individual level, yet work as a unit to provide shelter in a highly prescribed environment like that on Mars."

There are more than twelve hundred species of nature's "miracle" building material, bamboo. Members of the grass family (as are palm trees), bamboos are incredibly strong while being super lightweight. There are species of lumber bamboo that grow three feet per day, to a height of 150 feet. Anyone who has visited India or Asia has no doubt noticed the bamboo scaffolding attached to the sides of towering skyscrapers. After World War II, Philippine researchers constructed skins for covering wings and fuselages of high performance light aircraft from bamboo. The applications for this underutilized, renewable, and beautiful resource are far ranging. The shoots are edible and the stalks make quality paper. Thomas Edison's first light bulb factory in 1882 even used filaments of bamboo. We clear-cut forests and destroy entire habitats to produce paper when superior, cheaper, and far less destructive bamboo alternatives are available.

The history of human uses of bamboo provides a compelling example of early biomimetic innovation. Watercraft have been built of bundled reeds, grasses, or bamboo since prehistory. These bundles were typically thick and wide in the middle to create a serviceable raft or hull. The ends of the bundles were thinned out and drawn together to a point and then curved upward to provide a streamlined and seaworthy entry to the water.

The great innovators of the ancient world, the Chinese, went one better. They had used the extraordinary qualities of bamboo for a vast array of applications, from fishing poles to brooms to windmills and cranes. They studied the structural properties of a single piece of bamboo and concluded that they could be mimicked in the design and construction of boats of enormous size.

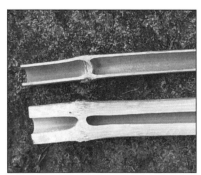

From left to right: bamboo cross-section, bamboo stand

The junk was born. By using bamboo to build masts and spars, as well as copying the internal structure of a bamboo stalk, the Chinese were able to stretch their scarce wood supply and build very lightweight craft up to a football field in size. The hull of a junk has strength-giving, watertight bulkheads, just like inside the stalk of a bamboo. By separating their hull sections with dividers, junks were more rigid and less vulnerable to sinking than any other designs. This was two thousand years before it became standard shipbuilding practice in other countries— and it wasn't until the designer of the *Titanic* "invented" similar bulkheads in the twentieth century that the technique entered the modern age. Unfortunately, unlike bamboo and the Chinese adaptation to junks, the *Titanic* only had partial bulkheads that did not seal the various hull segments completely. Had it been true to the concept, *Titanic* probably would have avoided disaster.

So capable was the junk that it has been the most successful ship design in history—a total of millions of ships being built continuously over two millennia. An archaeological dig near Nanjing has unearthed two enormous junk rudders that scientists estimate would have been fitted to a ship 450 feet long. No other wooden ship design could have been built to this size without breaking up. Christopher Columbus' *Santa Maria* was just 60 feet in length, Admiral Nelson's mighty *Victory* was 227 feet, and the U.S. clipper *Cutty Sark* 212 feet.

Fiber—it's good for the digestion and no tree or animal could stand without it. It's also a critical resource for many manufactured items, including any wood product, paper, clothing, rope, fiberglass, or composite materials. For millennia, humans have used organic fibers from grasses, palm fronds, silk, cotton, linen, wool, and even asbestos rock to create fabrics. Technologists are constantly looking to invent stronger, lighter, and cheaper fibers. A lighter car, plane, or boat uses less fuel, and a lighter soldier's helmet and backpack take less effort to carry. Thanks to industrial chemists, a wide variety of synthetic fibers have been developed to address thousands of applications. Essentially, all synthetic fibers are biomimetic, at least in copying the basic structure, but scientists are using advanced laboratory tools to study nature's secrets more deeply— exactly how nature bundles fibers and positions them in relationship to one another to achieve unmatched benefits.

During and after World War I, many aircraft were built of fibrous wooden frames covered with painted paper fiber for their outer skins. Now the U.S. Air Force builds stealth aircraft from carbon fiber and plastic resins—as does Boeing with their new Dreamliner passenger plane. Who would have imagined thirty years ago that we would be flying around in plastic planes? The annual global market for glass fiber alone is projected to be $16 billion by 2016.

Researchers Dr. David Hepworth and Dr. Eric Whale combined the best features of organic and synthetic fibers with a new biomimetic product, called Curran, which they offer as a substitute for glass fiber and hard to get, expensive, carbon fiber. The surprise? Curran is made of the nanofibers found in carrots. It turns out that carrot fibers are strong, stiff, tough, durable, environmentally friendly, and cheap. Hepworth and Whale's Scottish company, CelluComp, already produces high-performance fishing rods from carrots. That may sound like a small niche in which to start, but fishing rods are reported to make up the largest slice of the $20 billion worth of recreational fishing equipment sold each year. CelluComp has its sights on many other sporting goods, such as hockey sticks, skateboards, and tennis rackets, as stepping stones to even greater opportunities. This is an interesting business strategy, to sidestep what might be more obvious, larger markets, and establish credibility and market traction in a lower-risk arena such as sporting goods. After all, while an avid angler would be disappointed if his or her rod broke while reeling in a big fish, there's a lot less resistance to buying an innovative and potentially more resilient fishing rod than to suggesting that someone take the first flight in a plane made from carrots.

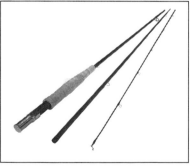

CelluComp fishing rod

Chapter 8

. . . .

WAMPUM

I had never been hit in the head with a duck's egg before. It felt more like a rock, except for the wet ooze that spread over my ear, neck, and shoulder. Though the same shape as a hen's egg, the duck-made variety is larger and decidedly tougher.

As the only Fisheries inspector in a remote coastal town of bored, off-season fishermen, it didn't take much for someone's imagination to dream up "Egg fight with the Fisheries!" The one store in town quickly sold out of eggs, and new supplies were three days away by truck. The fishermen had raided their fridges, pooled their egg resources, and launched numerous attacks against Fortress Fisheries—loosely including the government house/office, patrol vehicle, and my person.

Eggs are wonderfully and optimally streamlined and can be thrown with considerable accuracy over a long range. I, of course, in the interest of self-defense and Fisheries honor, was forced to retaliate. As the owner of the only live fowl in town (ducks), I had a steady supply of ammunition—though waiting for fresh deliveries tested my patience.

Although dozens of eggs had flown and had made their marks on house and vehicle, none hit me—until the last day of the battle. The fishermen had exhausted the town supply. I endeavored not to appear too smug and thought that the affair had ended. Just in case, I still carried a retaliatory egg in my hand everywhere I went. One of my foes, however, had snuck over my fence and helped himself to the sole duck egg of the day. I was walking across my front lawn to get in my truck when—

thwack!—that awful sensation. A loud cheer went up from a group of fishermen standing at the open door of a bar across the street. Without hesitating, I spun around and flung my own egg at them. In the time it took to fly across the street, the men scattered, the egg flew through the door—and smashed itself on the mirror behind the bar. Quite a mess I had to clean up. It looked, to all, like certain defeat of the Fisheries, but I had other tricks up my sleeve. However, that's another story.

It seems that whenever humans create technology to solve a problem, we end up causing detrimental side effects. We created the internal combustion engine, which is less than 30 percent energy efficient, and gave rise to asbestos, chromium, lead, and methyl tertiary butyl ether (MTBE) contamination of air, lakes, and aquifers, and escalating asthma rates in our children—to name just a few consequences. We created wood-pulp based paper and are subsequently responsible for clear-felled forests, erosion, spoiled rivers, collapsing ecosystems, and mercury poisoning. Plastics pollute with dioxins and mostly aren't biodegradable; thousands of new chemicals have never been tested for toxicity; and there's that most dangerous of substances—nuclear waste. On the other hand, when nature seeks to solve a problem, it resolves all associated problems.

Take, for instance, the humble egg. It's streamlined, so it can be laid with the minimum of discomfort to Mom. In fact, it's so streamlined that it's the shape a rectangular brick will be turned into by nature, if left long enough in a flowing stream. We all know how thin an eggshell is. The hen uses the least amount of energy and materials in its creation so as not to deplete her own calcium levels. If she can't replace the calcium needed to make eggshell quickly enough through her diet, she extracts it from her skeleton, causing a condition known as avian osteoporosis, and even her own demise. While fragile enough to break from the inside when pecked at by a delicate, hatching, chick, the shell is incredibly strong under compression from the outside. In 2001, scientists duct-taped a raw hen's egg to the side of a deep-sea submarine. It descended one thousand meters below the surface of the ocean—a depth that would crush almost anything. The egg returned to the surface intact.

Eggs are also highly thermally efficient. When Mrs. Hen needs to take a walk for food or other necessities, the egg holds and recirculates its heat efficiently so that it doesn't cool down too much during her absence. The egg is also optimized for nutrient distribution. Both its heat and nutrients flow in a highly efficient, doughnut-shaped ring vortex, perfectly matching the shape of the egg. Many eggs are elongated at one end. This out-of-round shape means that should an egg roll out of its nest, it will tend to roll back toward it again—handy if you've laid your offering on a cliff ledge. If the egg does take a fall, it has a shock-absorbing air pocket at the larger end. This also doubles as a pressure-relief valve if the egg gets overheated, since the air pocket compresses and protects the shell from expansion forces. The shell also offers the dual features of keeping out dirt and water, while being microscopically porous so that the developing chick gets oxygen as needed. Clearly, a number of adaptive strategies are related to an egg's shape—which matches the proportions of nature's spiraling whirlpools. These proportions can result in an infinite number of variations; hence we see many varieties of egg shape and size that all share similar benefits. Beyond all these attributes, the egg is entirely recyclable and leaves no deleterious residue.

Eggshells, seashells, crab shells, coral, skeletons . . . all these structures protect and support soft tissue. Whether inside, outside, around, or built inside and then squeezed out, like an egg, they're all created by living organisms that lay down layers of stone—calcium carbonate. Over the four hundred million years that these animals have been in the fossil record, massive walls of discarded shells have built the famous white cliffs of Dover in England and the marble mountains of Carrara in Italy. Fully 4 percent of the earth's crust is made of calcium carbonate, including limestone, chalk, and coral, 99 percent of which was formed from the skeletons of living organisms.

The phylum that includes the most shelled animals is Mollusca. Mollusks make up one quarter of all identified marine organisms, with more than eighty-five thousand species described; of these, seventy thousand are snails and slugs, with just a minority land dwelling. Most mollusks grow shells for protection, though a number of species use other survival strategies as well.

The shells with which nature equips many sea mollusk species have been of major importance to humans. Their inhabitants have been used as food, while their beautiful, compelling shells have served as jewelry, musical instruments, and sacred objects. Shell beads, called wampum, were woven into belts for ceremonial use and history keeping by Native American tribes of the Northeast. They were used as currency by European settlers and honored as a form of money in certain areas until the early eighteenth century. Murex sea snails were the source of Tyrian purple dye, prized at least as far back as 400 BCE, and so costly to make that it was worn only by the superrich of the time. It took twelve thousand snails to dye one toga. And until the advent of plastic, shells were the leading source of material for buttons around the world.

Now biomimics are seeing the beauty and value of seashells with new eyes, or in some cases, fingers. Geerat Vermeij is a tall, slim, professorial figure, which suits his position as a geology professor at the University of California, Davis. As an expert on seashells and the world's leading authority on the evolutionary adaptations of mollusks, Geerat travels the world to discover and examine their forms—with only one twist: He has been blind since the age of three. His love of their shapes and ability to feel the differences in their forms led to new insights into ancient mollusk history and allowed him to author a number of important books on evolutionary biology.

Seashells are masterpieces of design, function, and nanotechnology—earning the envy of the world's most skilled engineers. Humans cut flat lumber planks out of round trees, leaving piles of sawdust and offcuts—or extract and melt metal from rock and then grind, file, and cut it into desired shapes. Nanotechnology, on the other hand, is an inherently biomimetic emerging field that creates materials and machines by assembling them one molecule or even one atom at a time, with no waste. Nanotech spans disciplines as diverse as materials science, chemistry, molecular biology, physics, semiconductor engineering, and fabrication; the potential for creating efficient products and materials is truly fantastic.

With the benefit of nanophotography, scientists are now studying the way mollusks lay down their structures one molecule at a time, using the least material for the most effect. And fascinatingly, of the tens of thousands of spiraling seashell species, very few curve to the left. One is from India, where it's considered sacred, and another is a native of the waters near Florida.

Snails and slugs leave a trail of lubricating slime as they move over the landscape. The large yellow banana slug of the northwest United States exudes a slime that the local native people licked as a food source—each to his own. But not all slime is as delectable or wholesome as that of banana slugs. Australians, a rowdy, fun-loving lot, are often prone to pranks and dares. An Aussie recently was dared by his mates (Australian for *friends*) to swallow a garden-variety slug. Who knew that slugs feast on rat feces? Our gallant friend ended up in emergency care with a serious rat-borne disease.

You may not have noticed that snails and slugs move by galloping—they stretch the front of their bodies forward and then pull the rear up to meet it. This undulating, wavelike movement has inspired a biomimetic robot designed by researchers at the Biomechatronics Lab at Chuo University in Japan. The benefit is that the robot, like a snail, has a large part of its surface always in contact with the ground and is therefore very stable in each direction.

SHARP IDEAS

Limpet

Even the way a mollusk eats offers inspiring strategies. Many use a radula, which is like a ribbon covered with rows of many small, sharp teeth, to grab and scrape food into their mouths. Most radulas are built of fairly similar materials across species, though the common limpet, *Patella vulgata*, has magnetite and silicon dioxide cutting tips on its radula. Both of these minerals are supertough materials for rasping at its food. Magnetite has a hardness of approximately six out of ten on the Mohs' scale, which is the equivalent of your sharpest kitchen knives; and, silicon dioxide, known for its hardness since antiquity, is rated even higher at seven on the Mohs' scale. This means the limpet can get more eating done with less time and energy invested.

The chiton seashell is the product of about five hundred million years of research and development, and who knows how many prototypes. Although one of the most primitive of all mollusks, it is an exquisitely refined set of technologies. Researcher Derk Joester at Northwestern University in Illinois studies the chiton's self-sharpening teeth, which, as he told National Public Radio, are made from "one of the toughest and hardest materials known in nature." Chiton teeth, like seashells, are created in water at ambient temperature, so they don't require foundries with massive amounts of heat and pressure to manufacture. Chitons cling tenaciously to rock and grind it away with their teeth to extract their algae supper. That rock candy can be granite or quartz—harder than the steel of an ax. The promise of this research? Just for starters— superior dental implants and artificial bones for hip replacements.

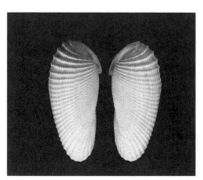

Angelwings seashell

While on the subject of toughness, the beautiful piddock, or angelwings seashell, looks fragile but its surfaces are so hard that the shell can bore into solid rock. By grinding back and forth, the up-to-seven-inch-long piddock can tunnel into stone, where it takes up permanent residence protected by a rock fortress. The $1 trillion mining industry could well learn something from this strategy.

Sea urchins are soft animals sheltered by a globelike shell and razor-sharp spines. They're echinoderms rather than mollusks, closer cousins to starfish than to sea snails, but they also munch on rock to make safe caves for protection from predators. Researchers P.U.P.A. Gilbert, Christopher Killian, and their colleagues at the University of Wisconsin in Madison have discovered that urchin teeth "are among the most complicated structures in the natural world." They're not only super-sharp but also made of interlocking, curved plates and fibers that are self-sharpening. Parts of the continuously growing tooth are designed to break away, leaving the remaining structure razor sharp—a little like chipping obsidian to make arrowheads. The researchers have confirmed that the sea urchin's strategy performs much better than man-made cutting tools. Such self-assembling nanotechnology could be used in industrial and construction tools to mimic the urchin's sharpening strategy.

Sea urchin

Researchers Helmut Cölfen and Jong Seto of the University of Konstanz in Germany are studying the extraordinary strength and fracture resistance of the spines themselves. As defensive weapons, they're necessarily very hard but also shock absorbing. This seeming contradiction has made them one of the most studied of all biomaterials. Urchin spines, although 92 percent calcite crystals, are bonded together with 8 percent calcium carbonate that has no crystal structure. The hope is to create much tougher, fracture-resistant biomimetic concrete.

Sea urchins are also a target of research on developmental biology, since their DNA is remarkably similar to humans. Dr. George Weinstock is one of the scientists who has studied the sea urchin's genome. Now at Washington University, his research confirms that humans share a common ancestor with these globular, spiny, little echinoderms—540 million years ago. In fact, the series of developmental steps that happen after a human egg is fertilized by a sperm turns out to be the same in sea urchins.

Dank and tepid, the smell fully filled my nostrils. I was immersed in it. I was in one of those all-encompassing assaults on the senses only made bearable by the knowledge that an unspoiled, living, natural environment produces this. Tiny sand flies clouded my face—falling into my eyes and breath. My arms and hands were too engaged pulling me through the water and mud to brush away flies and mosquitoes.

Am I nuts? I'm up to my chest in Crab Creek—a tidal, mangrove-choked channel through the foul smelling, muddy swamp of Roebuck Bay in the remote Kimberley region of Australia. I lost my shoes within the first steps, sucked off my feet as I pulled each leg up out of the greasy ooze. My downward descent only arrested when the mud reached my crotch.

"C'mon," encouraged my guide, the off-duty, local tow truck driver, famous for his routine of drinking a carton (twenty-four cans) of beer each day. At twenty-four years of age, his simple ambition was to increase by one, on each birthday, the number of cans he consumed daily. "Another hundred yards and we'll see crabs."

Mangrove crabs are remarkable animals. Their olive-colored bodies, the size of dinner plates only much tougher, ring like a bell when tapped. Their claws, each as thick as a man's forearm, are so powerful they can break the top off a beer bottle with ease—handy if you've misplaced your bottle opener. Their flesh is delicate and white and highly prized throughout Southeast Asia. From Sri Lanka to Indonesia and Australia, these Sherman tanks of the crustacean world live in burrows in the muddy banks of mangrove creeks and swamps. They also swim when not at home.

I struggled to keep my mind free of thoughts of what would happen to my most vulnerable parts if I agitated such a bottle crusher by sinking on top of him. Even after cooking, this crab's powerful claws and shell are so strong; it takes a hammer to crack them. Yet like an egg, their massive strength belies their relatively thin construction. It's all in the shape— nature's geometries of compound curves and proportions.

"There," whooped Ray. I looked and saw three holes in the muddy bank exposed by the receding tide. Ray had a steel rod with him—two feet long with a right-angle bend on one end. The bent section stuck out sideways about four inches. He slid and pulled himself through the slick mud toward the closest hole and plunged in the rod and his arm up to the elbow. I heard a clink, like steel striking glass.

"You beauty." Ray pulled hard and dragged out a flicking, snapping, and muddy ball of aggravated life. "Where's the bag?" he called out, with laserlike focus on his prey. I produced the jute sack I had wrestled through the creek with me.

"Watch this." Ray moved his right hand quickly toward the animal; the crab just as quickly lunged a massive claw at his hand. From outside the crab's range of vision, Ray used his left hand to catch the crab's other claw around its thickest part. Ray's feint had worked. He caught the first claw by moving from behind the crab's head and held the beast up in the air, a claw in each hand. "See how easy it is? Eight pounds, I'll bet." Ray was beaming. He dropped it into the bag I gingerly held open.

Five more crabs followed in quick succession—a full meal for six to nine people. They were particularly restless on our slipping, sliding, groping journey back to the Land Rover. I pulled the bag behind me—as far as possible from my body. My fingers ached as I clamped the bag's top in an iron grip. I wasn't leaving any possibility of a revenge raid by a carelessly liberated denizen.

CONCRETE SOLUTIONS

Brent Constantz, founder of the innovative cement manufacturer Calera, understands the chemical processes that a crab, a duck, or a mollusk uses to lay down its shell or bone. Brent—a PhD and professor at Stanford

University, a serial entrepreneur with several successful start-ups, and eighty-five granted patents—is focused on environmental sustainability. Although multidisciplined in his skills, his core expertise and qualifications are in biomineralization, and he has become an expert in developing medical cements for mending broken bones.

Did you know the biggest single cost in national health care is broken bones? And in particular, broken hips in postmenopausal women? As a research scientist and avid diver, Brent came to the conclusion that he could make bone fracture cement in the same way that nature grows seashells and coral. After all, they're made the same way as bone. Brent founded Norian Corporation in 1987 to develop a biomineralization technology. His fracture cement, uniquely, cured at body temperature rather than through heating up and therefore didn't cause damage to or destruction of living cells, tissue, or proteins. It proved highly promising when tested in initial human patients. There was no inflammatory or immunological response and no risk of disease transmission.

Norian skeletal repair system (SRS) received foreign governmental approvals in 1992, with U.S. approval following in 1996, and is found in most operating rooms that do orthopedic surgery today. Brent grew the company until 1998, commercializing the cement worldwide with several foreign subsidiaries, when he was replaced as CEO. "Unfortunately, I was burned by venture capitalists and a large corporate partner, Synthes," says Brent. In 1999, his rapidly growing company was acquired by Synthes Holding AG of Switzerland.

After leaving Norian, Brent founded a cardiovascular device company Córazon Technologies, which was dedicated to addressing the leading cause of amputation in humans, calcification of the cardiovascular system. He raised significant venture capital and was in clinical trials for revascularizing calcified arteries in patients who would otherwise have needed a leg amputated. As Córazon grew, "The venture capitalist that invested in the company took control and attempted a near-term commercial path that was unrealistic. It disabled the company," said Brent. Córazon was eventually acquired by Johnson & Johnson, which is now commercializing the technology.

Meanwhile, some former colleagues from Norian convinced Brent to form a new company to develop a better bone cement than Norian. Skeletal Kinetics was founded in 2001, and the experienced team had their new cement in operating rooms in less than a year. This time Brent didn't raise venture capital funding. As he describes, "There's the real world and then there's Silicon Valley. When you take venture capital money, you need to understand: The deal is set up for the VC, not for you." He adds that in his experience, "Overfocus on directors can debilitate companies. Management wastes up to 20 percent of its time preparing reports for venture capital board meetings and decompressing afterward. Most venture capital partners are on more than one board and many turn up to board meetings for their start-up companies without having read the carefully prepared reports. Uninformed, they then issue a shoot-from-the-hip analysis and instruction that changes by the next board meeting." Brent reports that without the burden of a large board of directors or venture capital backing, he thoroughly enjoyed his work with Skeletal Kinetics.

Meanwhile, Norian, now part of Synthes, had developed a new version of the Norian SRS cement without understanding its biomimetic origins. As Brent explains: "Synthes added barium sulfate to a cement that was otherwise identical in chemistry and crystallography to natural bone. Synthes thought this would make the product more appealing to surgeons by being easier to see on radiographs and fluoroscopes, completely disregarding its biomimetic design. And then—catastrophe. Putting profits before caution about patient safety, Synthes proceeded to test the new products on approximately two hundred patients, without FDA approval and after the FDA specifically warned against their use. Terribly, three patients, all members of military veterans' families, died. Instead of stopping development immediately, Synthes executives covered up, and false statements were made to the FDA. The Department of Justice was outraged. Synthes eventually pleaded guilty to a number of crimes and was fined the maximum of $23 million; several senior executives, including the company's president, were indicted."

This is a tragic example of corporate greed and lack of integrity damaging a valuable biomimetic technology.

A turning point in Brent's life came when his son succumbed to cancer in June 2000. It became an even stronger imperative for Brent to contribute to the protection and health of the natural environment and he accepted a position as an associate professor at Stanford University. As he advises students in his lectures, their own children will probably never see live coral reefs, because reefs worldwide are dying so rapidly. Coral is especially important to Brent, since it inspired him to invent not only a means to synthesize coral skeletons in human bones but a whole new way of making construction cement.

Coral is built by trillions of tiny organisms called polyps that extract magnesium, calcium, and carbon from ocean water to build a community of skeletons. In a similar way, Calera, which Brent founded in 2007, creates cement by running carbon dioxide from flue gas from a nearby power plant through water containing calcium, magnesium, sodium, and chloride—such as seawater This combination of chemicals and minerals converts the carbon dioxide in the flue gas to related materials called carbonates, which are heavier and precipitate out of the salt water. After removing the water and drying, the product is ready for use as cement. In effect, the company makes chalk, and indeed, Calera's cement is bright white. As a side benefit, the source water has had its salt removed and can be purified to fresh water with only a few additional steps. Finally, and perhaps most significantly, rather than giving off carbon dioxide to the atmosphere, Calera's process absorbs half a ton of carbon dioxide for every ton of cement it produces.

There are four hundred billion tons of carbon dioxide in earth's atmosphere, with billions more added each year due to human activities—and burning fossil fuels is not the only major source of humans' contribution. The release of carbon dioxide in the process of cement making for concrete is a leading source, because one ton of traditionally produced cement generates at least one ton of carbon dioxide. As concrete is the most traded material in the world after water, cement production has become the third largest contributor of carbon dioxide to the atmosphere, at three billion tons per year.

As Brent says, "We think, since we're making the cement out of carbon dioxide, the more you use, the better," says Brent. "Make that wall five feet thick, sequester carbon dioxide, and be cooler in summer, warmer in winter, and more seismically stable."

Calera can also make the crushed rock aggregate used in concrete. Presently, the world mines forty trillion tons of aggregate per year at a huge energy and environmental cost; 60 percent of the aggregate used in U.S. concrete is mined in British Columbia, Canada, and shipped south. The potential reduction in additional carbon dioxide emissions through manufacturing aggregate closer to a job site, rather than mining and transporting it, can be calculated in billions.

Are there alternatives to Calera's process? There are early programs under way to capture carbon dioxide from flue gas and transport it underground, with the intent of storing it permanently in the earth substrate. There are a few problems. The cost of trapping, separating, transporting, and storing all that carbon dioxide could be as much as a third of the energy output of a power plant. And if an underground vault leaks through earthquake activity and the carbon dioxide finds its way back into the atmosphere—a total waste of effort and money.

The scope of Calera was so large that Brent felt he had no choice but to include venture capital funding. The company's technology and market was so compelling that it was able to raise $200 million over three years, without a formal business plan. Brent joked that the main people who read business plans are your competitors, and the lofty scenarios outlined in them are often the only things that don't happen to a young and growing company. Flush with cash, Calera expanded staff at the rate of forty per month and built a two-hundred-acre pilot production plant.

Calera was seen as one of Khosla Ventures' most promising portfolio companies. It grew rapidly to a staff of 140, with a mandate from Vinod Khosla, the senior partner of Khosla Ventures, to hire the best people the world had to offer.

"With large venture capital backing, you can hire anyone," confirms Brent. "As well as the multiple challenges any start-up faces in proving its technology and scaling it for market, there was plenty of criticism, lots of politics, and no shortage of disinformation circulated by competitors and small thinkers. There was also huge venture capital pressure to make a lot of money very quickly."

In late 2010, the board of Calera replaced Brent as CEO with the former CEO of a Canadian mining company. Brent concludes, "It was very stressful and disappointing. I had just raised one hundred twenty-five million dollars for further development; they haven't raised any money since I left. As of late 2011, the company had spent all but $50 million of its money. Staff was reduced from 140 to 40 and it is unclear what, if any, commercial deals the company is undertaking."

Will Calera make it? Certainly nature elegantly achieves what Calera is trying to mimic. Brent and his research colleagues are confident in the technology. As he says, "Calera absolutely deserves to work. It offers a huge and unmatched potential to sequester a large percentage of the world's carbon dioxide emissions." To be a market success, the company needs to be able to scale its production to commercial levels while applying the right business model for international growth—a complex challenge for any technology. The real issue, according to Brent, is a combination of cynicism and outright hostility from the status quo, which is quite typical with new paradigm-changing technologies, and the efficacy or otherwise of management, funders, and commercialization tactics. "The biggest obstacles we face in this world are doubt and greed," says Brent. "The world needs this technology. It's whether it happens now or in thirty years. It's just a matter of time before we're faithfully copying nature and creating benign cement."

I asked Brent if he had any tips for other aspiring entrepreneurs. "Do something that has significance," he replied, "something that makes a difference and that you have passion for. Second, successful entrepreneurs are the ones who follow through. As Winston Churchill said, 'Never, never, never, give up.' Third, don't set goals too high. Fourth, do it without raising lots of capital. Fifth, stay focused. Sixth, every quarter, make an operating plan, and stick to it."

He emphasizes the point that, in this rapidly changing world, the biggest problem is deciding on which markets to focus your effort. For instance, the economies that dominated the world in the twentieth century—the United States, Europe, Japan—are all contracting and being supplanted by former Third World countries like Brazil, China, and India. The demographics of these countries are important indicators. In India 65 percent of the population is under thirty-five years of age, and 31 percent is under fourteen years old. Brazil has 27 percent of its population under fourteen. China, on the other hand, with the world's largest population and second largest economy, has only 16 percent under age fourteen. In addition, Brent points out that although China has a powerful appetite for growth and plenty of capital, thus far it's difficult for Western companies to make money there. What does all this mean to market opportunities in the coming thirty-five years? A successful entrepreneur, biomimetic or otherwise, must think strategically, not just about the particulars of his or her own technology but how global markets—delicate ecosystems in themselves—are connected.

SHUCKS

Steamed mussels

Calera is a compelling example of the huge commercial potential for biomimicry, but there's even more that we can learn from shells. Mussels, the shellfish that we order in Italian restaurants, don't just taste great; they offer sustainable chemistry at its best. A subgenus of mollusks, mussels cling to underwater posts and rocks using a nontoxic, biodegradable organic resin or glue that is at least as strong as any adhesive invented by humans. It sets up under water, yet can be dissolved without introducing any solvents. Mussel adhesive proteins, or MAPs, have interested researchers for years. Thanks to a number of university studies on the structure of their adhesive threads and chemical features, teams around the world are now moving toward or have achieved commercialization of MAP products.

Kollodis BioSciences of Massachusetts has developed a proprietary method for scaling-up the manufacture of MAP adhesives. While still a relatively small firm, the company has full manufacturing facilities and sells products through their website. Initial products for the scientific research industry include a three-dimensional cell culture matrix and coatings. These act to modify plastic surfaces and membranes in order to promote the attachment and growth of cell samples that doctors and lab technicians need to count, examine, or treat.

A synthesized form of mussel glue is also showing tremendous promise in medical applications, including the coating of medical devices, since the combination of chemicals in the sealant is biocompatible and doesn't trigger rejection by the body. Phillip Messersmith of the Messersmith Research Group at Northwestern University, with colleagues from Canada, Belgium, and Switzerland, have completed laboratory tests and are looking toward clinical testing.

Mussel genius doesn't end with biomedical applications. Much of the world's forests cut down in the past fifty years has been used for plywood and oriented strand board (OSB), including 99 million cubic meters just in 2010. Remember the hundreds of emergency trailer homes that the U.S. federal government supplied to Hurricane Katrina victims? Large numbers of them were never used, because the plywood and chipboard they were made from outgassed so much pungent, toxic formaldehyde that they were declared to be health hazards. Formaldehyde causes eye, nose, and throat irritation as well as breathing problems and even asthma, vomiting, and severe headaches. It is also a known carcinogen.

Columbia Forest Products has now developed PureBond, a formaldehyde-free plywood. As the company's website describes, Dr. Kaichang Li of Oregon State University found that "mussels secrete proteins known as byssal threads." These are superstrong, flexible threads that give mussels their ability to hang on to rocks even in the wildest storms and wave movement. Dr. Li was able to modify nontoxic soy proteins to create similar properties to byssal threads. An added bonus to their unmatched strength and nontoxicity is that they're waterproof. Columbia Forest Products won the EPA's Greener Synthetic Pathways Challenge with this nontoxic, LEED (Leadership in Energy and Environmental Design) and CARB (California Air Resources Board) compliant, and very importantly—cost-competitive—product.

DEEP AND MEANINGFUL

You've probably heard how little we know about the ocean depths and their inhabitants. One snail found living near sulphur-spewing hydrothermal vents, thousands of feet down in the Indian Ocean, surprised researchers. Many mollusks have a defensive front door for their shells called an operculum. Our deep ocean snail, affectionately called scaly foot, has a unique operculum made of golden-colored scales reinforced with layers of fool's gold (iron pyrite) and gregite (a mineral used for sparking flintlock guns in the sixteenth century). No other animal on earth is known to incorporate fool's gold or gregite. Incidentally, these materials are highly magnetic, as is the magnetite in the earlier-mentioned common limpet. Nature doesn't normally invest in coincidences, so why do these mollusks use magnetic materials?

Although scientists haven't yet answered this question, these highly evolved strategies are already offering value for human adaptation. In early 2010, the MIT news office reported that Dr. Christine Ortiz, Dr. Subra Suresh (then Dean of Engineering and now head of the National Science Foundation), and their colleagues used a diamond-tipped measurement tool, as well as computer modeling, to analyze the mechanical properties of the scaly foot's shell. They discovered that "its unique, three-layered, structure dissipates mechanical energy, which helps the snails fend off attacks from crabs" who try to break the shell with their claws (which are in themselves a phenomenon of strength and leverage that can help us optimize the design of wrenches and levers). "The shell of the scaly foot possesses a number of additional energy-dissipation features compared to typical mollusk shells that are primarily composed of calcium carbonate." The industrial opportunities already anticipated include superior helmets, protective armor, and new structural materials. Pyrites and gregite are also being evaluated as an alternative to silicone for the creation of cheap, abundant solar cells.

Operculum is Latin for "little lid" and biologists have assumed that this hard trapdoor provides its species with protection, prevention of water loss while exposed to air when the tide is out, assistance with locomotion, and perhaps some form of offensive measure. But, since an operculum also looks so much like an ear, my view is that it's likely it performs a similar function to ears.

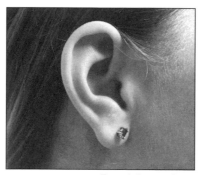

From left to right: ear, seashell operculum

Hearing works on pressure waves. Seashells and fish detect one another, and danger, through pressure waves. The incompressible nature of water means that pressure waves travel very long distances without dissipating. By means of this phenomenon, pods of whales are able to communicate with one another other even when several thousand miles apart. Navies monitor one anothers' ships on the other side of an ocean in a similar way. It's not a stretch to theorize that, just as the specific spiral geometry of an ear and its cochlea help to catch, channel, and amplify sound waves, the same ear-shaped geometry of an operculum channels pressure waves for a mollusk. Handy really, since the animal, like any snail, is vulnerable when outside its protective shell. An operculum sits outside the shell, right on top of the animal's foot when it's walking, and would make an excellent early warning system, so that the mollusk could withdraw to safety inside his shell at the sound of danger. In fact, I speculate that the entire outer shell in many species might double up as a pressure-sensing technology. If true, these geometries may improve efficiencies in antenna, speakers, microphones, and transmitters.

If you're not thrilled at the prospect of having multiple poisonous harpoons fired into your hand, don't pick up a cone shell you find on the beach—no matter how pretty. Some of the six hundred species sting like bees; others can be fatal—and there is no antivenin. The components of cone shell snails' venom act with high speed and precision. In particular, they can target a particular class of chemical receptor to the exclusion of any other. This makes them ideal to synthesize for use in instant reduction of heart rates or turning off pain receptors.

Cone shells

A particularly potent species of cone, the *Conus magus*, was analyzed in the 1980s and resulted in the development of Ziconitide, which is sold under the name Prialt. By "blocking the calcium channels of the nerves that transmit pain in the body," it is actually one thousand times more powerful than morphine. Approved for use in 2004, it's not addictive like morphine. However, it does come with serious side effects, so Prialt is only used for the treatment of chronic, severe pain that no longer responds to other treatment. Other compounds isolated in cone venom show promise in the treatment of epilepsy, Alzheimer's, and Parkinson's disease.

SKINNY-DIPPING

Some classes of mollusk do just fine without a shell. Nudibranchs, a remarkable and shell-less subset of the mollusks, have highly evolved methods of communication and protection that offer surprising opportunities for bio-inspired design. The three thousand or so species include the most colorful creatures on earth (or rather, in the sea). They're exquisite in decoration with a huge variety of forms. One species, called the Spanish dancer, sports beautiful scarlet and white ruffles that look just like a dancer's flounced dress. This shell-less sea slug twists and gyrates its entire body in a choreography of movement that propels it along.

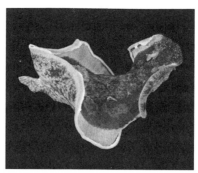

Spanish dancer nudibranch

As Nan Criqui of the University of California San Diego's Scripps Institute of Oceanography describes, scientists have discovered that these mollusks are a veritable chemistry lab, storing and recycling what they eat to great advantage. Their brilliant coloration and toxicity to predators results from reusing chemicals in their food. In particular, nudibranchs have a penchant for snacking on stinging sea anemones. While it was originally thought that a layer of mucus protected them from getting stung, researchers instead discovered that nudibranchs have evolved a storehouse of different chemicals—specific to each type of anemone—that neutralize the prey's poison. Nudibranchs go one step further and absorb those toxins so that they can be reused in self-defense! Anesthesiologists are interested in better understanding the chemical mechanisms underlying this strategy so they can improve pain control in patients. To humans, nudibranchs can smell like "lemons, watermelon, vanilla, or flowers." Researchers are identifying the chemicals that trigger those smells, as well as the chemical messages about danger, mating, or food that their slime trails leave hanging in the water. These signals are laid down in a mucus that stays put even under water and act as instant messages and invitations to the next passing nudibranch—sort of like slime mail. Potential biomimetic applications include the inhibition of unwanted chemical signals given off by medical and industrial chemical processes.

One thing I am particularly struck by is the apparent intelligence of what are usually considered dumb animals. For instance, if you place food morsels into a screw-top jar and place the jar near an octopus, the animal will work out how to unscrew the lid to access the goodies. I know of an octopus at a celebrated aquarium that would wait until the attendants left for the night, climb out of his tank, walk across the floor to adjoining fish tanks, eat his fill of their occupants and then return to his own tank before the staff arrived in the morning. It took bewildered workers some time and an infrared surveillance camera to figure out why their fish were disappearing.

Octopi, squid, nautilus, and cuttlefish are considered to be the smartest of all invertebrates. These spineless cephalopods are another class of mollusk and are cousins to the gastropods, which include sea and land snails. *Cephalopod* means "head-foot," since these animals are mostly head and lots of legs that developed in the place of a snail's single squishy foot. Some still live in shells, like the nautilus, while others carry a remnant of a shell inside their bodies (the cuttlefish) or none at all (octopus).

Squid, octopi, and cuttlefish invented the equivalent of polarized sunglasses eons before us. Dr. T. H. Waterman of Yale University's Osborn Zoological Laboratory discovered in 1950 that cephalopods and many fish, crustaceans, arthropods, and aquatic insects have specialized mechanisms in their eyes that analyze light polarization. This ability lets them significantly reduce glare from bright sunlight. Dr. Waterman continued his research for decades, though the developers of Polaroid sunglasses, which were originally invented in 1936, undoubtedly had no idea that they were biomimics of marine animals. We could learn more from these highly adapted creatures, and the researchers who have already studied them, in order to optimize our modern optics, telecommunications, and medical devices.

As we've seen, even a common garden snail can teach us about nanotechnology, optimized balance, minimization of materials, streamlining, defense, camouflage, and aesthetics. Who knows what we'll discover in coming years, as scientists zoom farther into the depths of the oceans and the molecules of mollusks. But you don't need to be a biomineralization expert, a zoologist, or an engineer to discover a valuable biomimetic solution to formerly intractable problems. Any of us can discover and develop something new if we keep our eyes and curiosity open. The treasure chest of nature's secrets is waiting for you.

THE NATURE OF CHANGE

Chapter 9

. . . .

THE CORPORATE JUNGLE

What are the best business models for bio-inspired design?

Should we patent inventions based on what we learn from plants and animals or use an open source model?

Should we license innovative ideas to existing manufacturers or build products and sell them?

How do entrepreneurs, biologists, and engineers break age-old traditions and overcome the obstacles to success?

It was closing time at Sydney's largest convention center, where I and hundreds of others had spent two days setting up displays for the international boat show that was opening in the morning. My chest was thumping and I felt more than a little nauseated as I gave my booth one last look over. My budget for everything was just $20,000, so at 250 square feet, it was one of the smallest available—less than one-twentieth the size of Yamaha's pavilion. The huge Japanese boat and motor maker was a main sponsor of the event and always invested major resources into their displays. Everyone expected Yamaha to win first prize for its presentation.

I was told I was lucky to have even gotten a space, particularly since I was a total newcomer to the industry and my novel biomimetic boat design was wacky. I had to agree—at least in comparison to the other boats at the show. My years of being in or on the ocean had given me a great appreciation for the efficiency and elegance of marine animals, but their shapes weren't used in the boating industry. WildThing, as we named it, was modeled on the curving forms of orcas and dolphins. It had no heavy ribs or frames in its construction and was designed like a flying fish, to quickly lift and skim across the top of the water. All this added up to an unusual, organic-looking craft.

If leaving the convention center that night was unsettling, driving to it the next morning was worse. Here I was, trying to break into the ultra-conservative, traditional boating world with a total of three employees—none of whom had any background in the industry—and a design that was unconventional in concept, shape, performance, and materials. I was pretty well known due to my previous businesses, so the press was going to be closely scrutinizing "Jay Harman's latest idea." I wondered if my biomimetic design would make me a laughingstock.

My ruminations were interrupted as we arrived and were swept into activity. The booth was surprisingly busy—more people coming through than we'd even hoped for. That day and the next flew past as people stopped to ask questions, take photos, and run their hands over our display boats. More than a few children and adults tried to climb in. With all the traffic, I was a bit distracted on the closing afternoon of the show when the head of the prize committee came to the booth and asked to speak with me privately.

"We have something of a problem," he said quietly. As he explained, Yamaha was scheduled to get first prize for all the money and effort it had put into its elaborate display and the show overall. But a head count and survey of attendees showed that our booth had received more visitors than any other and was considered to be the most original and compelling by far. To resolve the dilemma, he explained, the show's organizers had decided to give two, equal, first prizes that year. What a relief and thrill—WildThing was off and running.

THE THREE PRINCIPLES OF RUNNING A BIO-INSPIRED BUSINESS

Over the years, I've found three key ways that a bio-inspired business differs from other start-up companies. First, biomimicry by its nature is a cross-disciplinary process. There's an obvious reason for this, because biologists are not usually involved in building machines and engineers don't usually dissect heart muscles or collect rattlesnakes. Biomimicry moves problem-solving strategies across scientific disciplines, such as from marine biology to mechanical and aerospace engineering, zoology to materials sciences, or ecology to agricultural engineering. Yet each of these fields of study has its own metrics, shorthand, periodicals, and respected researchers. The world of business, with its processes for operations, marketing, and finance, is the same.

The last booms of electronics hardware and software occurred when innovative engineers emerged from existing engineering companies or military backgrounds. They were already familiar with the language and processes of their manufacturers and clients, and launched a new industry from the platform of an older one. By comparison, innovation based on nature often starts from outside traditional engineering or business settings. Enter a passionate scientist who has breakthrough insights on the biomechanics of a beetle or the chemical processes of photosynthesis but no experience with either product development and manufacturing or corporate governance and business operations—and you have different cultures at work. Obviously, there are translation issues that biomimetic business leaders must deal with in order to *cross the divide*.

A second core difference I've found is that bio-inspired businesses encounter particularly high inertia. Biomimics, generally, are extrapassionate. It's more than a job for them and more than a means to riches— it's a global emergency. They're not aiming to sell the next Pet Rock; they're trying to reduce or reverse planet-threatening damage caused by existing industrial practices or solve intractable health issues for humanity. At the other end of the scale, like large ships steaming ahead on a set course, are those established traditional industries that the biomimic wants to board. Entrenched interests with big market share and engineers with careers built on old paradigms can put up stiff resistance. Whether highly regulated pharmaceutical companies or automobile manufacturers trying to stay afloat, the traditional corporate mandate is to avoid risk and play it safe. A biomimetic leader has to *turn the hulking ship*.

Third, growing bio-inspired products is more like investing in an orchard than planting a crop of tomatoes. Many biomimetic products, though they can be at least as profitable in the long term, will take longer to deploy than the online services and innovations of the high-tech world. Unlike the computer software boom, biomimicry is more often reengineering nuts and bolts and molecules, rather than bits and bytes. We simply can't install bio-inspired solar panels on thousands of roofs, run clinical trials for FDA approval, or build energy-efficient skyscrapers on a dot-com timeline. In addition, scaling biomimetic processes up to full production sometimes requires different techniques than those offered by current mass production. Whether developing a project in an existing, larger company or in a new start-up, a biomimic must be realistic about resources, set expectations appropriately, and *plan for patience.*

Hearing this list of differences, which may at first sound like additional challenges to a new venture, you might wonder whether a bio-inspired venture is worth the effort. I can say with confidence that in terms of job and personal satisfaction, it is indeed worthwhile. We've stubbed our toes in a variety of ways over the years; sharing the lessons we've learned can hopefully make your path easier.

From small start-ups to stock market giants, a company's technology, team, and market are the three inextricably related aspects of any business. A bio-inspired technology must prove its competitive performance and economic viability against existing standards. Beyond that, it's a fact that a great team and the right market will have as much or more impact on your success as your technology. You must find the right people to support the project, both internally and externally, as well as being rigorously realistic about the ecosystem of your target market. Given this, how can you maximize growth and minimize delays? Let's look at how the three key differences of a bio-inspired business impact its technology, team, and market.

TECHNOLOGY—CROSS THE DIVIDE

Every innovation enters an existing industrial system with long-standing processes, from how to demonstrate proof of concept to prototyping to certifications to manufacture. Ignoring or trying to leapfrog standard technical requirements can hamper any project, but this holds particularly true for a biomimetic technology that is, most likely, out-of-the-box thinking to its engineering audience. Bio-inspired solutions usually represent not just an incremental change to an existing technology but also a total rethink of how to solve a problem. That gives tremendous potential for performance improvements, but there can be aspects of the improved performance that aren't yet fully understood or may even represent "new science." This can sometimes make hardened engineers nervous, since they've been professionally trained to eliminate risk by understanding all the implications of a design—both good and bad.

Given the first key difference between traditional start-ups and biomimetic companies—translation issues across disciplines—you can expedite progress by anticipating technical confusion and planning how to overcome it. Establish common language and goals in your own team. Don't assume that the MBAs, engineers, and biologists at the table really understand one another, because they probably don't. When in doubt, and even when not in doubt, define the key terms that are used across departments. Agree on one shared language by using the vocabulary and the metrics of the market you've decided to enter.

It's harder to build rapport with potential partners or clients when you have to translate your charts and graphs. In our early days at PAX, we generated test results that proved "higher efficiency" in our concept prototypes, but each of our target markets defined *efficiency* differently. Be easy for your client to deal with. Confirm that your performance graphs and reports provide data in a format that's familiar to them. Busy engineers or businesspeople in client companies are already new to your technology; they need to feel confident that you not only understand their metrics but also will be good communicators.

The old saying to not "suffer fools gladly" can't apply here. It's easy to be impatient with others who aren't immediately convinced, but that won't help you. All new technologies have to be sold to someone, whether a boss in an existing organization, a potential investor, a grant funder, or a client. Entrepreneurs and scientists—including extrapassionate biomimics in a hurry to change the world—can become so convinced of the merit of their ideas that they brush off hard questions and label anyone who doubts their technology as simply a naysayer. Be ruthlessly practical about the skeptics who you need to satisfy. If you want to win the game, you must know the rules. Dismissing expert opinions that disagree with yours or claiming results that can't be confirmed by outside testing is foolish. We found at PAX that working with existing institutions—commissioning research at universities or presenting papers at academic conferences and in scientific journals—is a useful practice to introduce our technology to experts in the field and gain credibility. If your concept has merit, it will stand the test of these peer-reviewed forums.

In the early days of PAX, we commissioned senior researchers at Stanford University's Center for Turbulence Research to test our theories. They listened to my description of how nature moves liquids and gases and watched my prototype demonstrations with interest, and then courteously tried to talk us out of spending our money on their technical analysis. A whiteboard was soon covered with numbers and Greek mathematical symbols that "proved" there should be no difference between my impeller shapes and conventional devices with the same surface area. They certainly knew much more about traditional approaches to fluid dynamics than I did, and I couldn't argue with their logic. However, I knew something about natural shapes. They are more efficient—period. I asked the scientists to humor me and I would accept whatever results they found. The team began their testing. When the first results came, three months later, and confirmed significant differences in the way our impellers interacted with fluid flow, one of the professors laughed with surprise as he said, "Who would have thought!"

Most conventional technology was and still is made of stiff, inflexible materials with large motors and few sensors. Nature's technologies, by contrast, are usually small, curved, and flexible—with many "sensors" that optimize performance under changing conditions. Despite many recent innovations in manufacturing, modern engineering is still based mostly on straight lines and ninety-degree turns, and some manufacturing engineers are resistant to trying to build the forms favored by nature. Anticipate their objections by bringing prototypes and third-party evaluations or endorsements that confirm that manufacturing your design is both viable and economical. Even when you prove your point, don't be surprised if there is still some skepticism.

In the early 1990s, I took my hand-drawn designs for the WildThing marine craft to one of the world's leading naval architects. I needed him to enter the design into a computer so that a computer-aided design (CAD) file could be generated to have production molds prepared for the boat. With no attempt to conceal his disbelief, he told me, "You're wasting your money and my time. If your design works, I'll eat my hat." I invited him to bring his knife and fork to the boat demonstration. We subsequently won international awards for the craft, showcased it in sixty countries via television features, and sold several thousand of them before selling the technology to a venture capital firm.

Product development is now a well-evolved practice worldwide. A series of stages and substages move an idea from concept to prototype to engineered device (or process or chemical) to preproduction sample to manufactured product. Rather than reinvent the wheel or use methods solely from scientific research (where many biomimics come from), there are excellent product development processes that biomimics can use to maximize their success.

PAX has a technology that impacts so many potential areas; it can be likened in some ways to a new pharmaceutical discovery. Because of that, we found that product development processes from the biotech sector were sometimes more helpful to us than those of mechanical engineering. Some homework will reveal the models that best serve your technology and market. The caveat is that your customers still need to see the reports and results of your product development delivered in their language.

TECHNOLOGY—TURN THE SHIP

Bringing a new technology to an old market can sometimes feel like swimming upstream in molasses. This inertia is at its peak in the massive size and complexity of the military-industrial complex, but one of the areas PAX initially identified for development was propellers and the navy seemed the natural place to start. Following an excellent first meeting with Rear Admiral Richard Riddell of the Office of Naval Research, I was introduced to one of the navy's senior scientists, Dr. Spiro Lekoudis, for follow-up. Although I had no interest in being involved with weapons programs, I have great respect for our military. My father, an Australian veteran of several tours in World War II, told me often that there would be no democracy on earth today if the United States had not entered World War II. I was willing to support bringing energy-efficient design to the greatest user of fossil fuel in the world—the U.S. military.

I learned at an energy briefing for a committee of the U.S. Joint Chiefs of Staff that 70 percent of all the weight of equipment on a modern battlefield is fuel and its protection and transport. For example, a forward base in Iraq was one hundred miles from the main fuel supply depot. Because of attacks from insurgents, road tanker convoys between the depot and the base were supported by aircraft and helicopters. Over the course of one year, it took two million gallons of fuel to protect and deliver one million gallons to the front line. For a highly fuel-dependent, mechanized army, energy efficiency is a matter of national security.

The U.S. military is also the world's largest funder of all manner of research.

Dr. Lekoudis listened to my concepts and reviewed my scale model prototypes. He advised that the navy was no longer conducting much advanced propeller research. For the navy to be able to assess our concepts, I should supply a twelve-foot-diameter propeller for them to test. A twelve-foot propeller would have to be cast in solid bronze, weigh thousands of pounds, and cost millions of dollars to design, computer model, build, test, and deliver. He also advised that even if we did supply a superior propeller, because of military procurement timelines, testing, and red tape, it would be at least ten years before it made its way onto a ship. He did, however, direct me to the Electric Boat Company (EBC)—a private contractor and a main builder and supplier of vessels to the navy.

I subsequently met with the good folks at EBC who, although courteous and interested, saw no way for my ideas, which were still in proof-of-concept stage, to fit their business model. They directed me to General Dynamics—builder of nuclear submarines, among other things. Upon arrival at their high-security facilities, my partner, Francesca, and I found ourselves in a maze of clearance requirements. Although the meeting had been arranged for weeks, it took more than two hours to make it past the front gate. We were finally met by a friendly chaperone who led us to a meeting room.

"There will be about a dozen people coming to hear your presentation," he said. "However, because of security protocols between departments, as well as your outside status, no one will introduce themselves to you. No one will ask questions, and you are not permitted to ask them any questions."

"What are their engineering specialties?" I asked, curious about whether I should address mechanical engineering specifications, fluid dynamics concepts, or structural engineering concerns.

"I can't answer that," was his reply.

We gave a forty-five-minute presentation to ten expressionless, silent faces assembled there and were then led back to the main gate. We were ushered through the steel and barbed wire entrance by armed security guards and found ourselves standing on the street with no feedback of any kind.

Some weeks later we received a phone call from the Office of Naval Research, asking us to come to a special meeting of researchers in Washington, DC. We were told there would be navy staff and contracted researchers from Pennsylvania State University in attendance. We were also advised that we would need to pay our own way. The gentleman on the navy end of the phone ended our conversation rhetorically with, "What's this PAX stuff anyway? *Pax* means peace and we're warriors."

"We'd like to help you keep the peace," I replied.

At last, I thought, we'll meet the right contacts and enter this complex ecosystem. When we arrived at the hotel conference room that was the appointed meeting site, I opened the door to see about ten people sitting around a large table. An individual jumped up from his chair and hurriedly ushered us out of the doorway back into the hall.

"Your presence is no longer required," he advised.

"What do you mean?" I asked.

"We don't need you to attend," he replied.

"You're kidding, right?" I was incredulous. "We were invited here, paid our way, and now you dismiss us with no notice? How is that reasonable?"

"Wait here," the man instructed. He turned back into the meeting room. A few minutes later, the door opened again and he reappeared, offering us fifteen minutes to address the group.

About five minutes into my presentation, a senior-looking fellow interrupted to ask, "How did you get here?"

"We were invited to make a presentation," I responded.

"Okay, you can leave now. We'll be in touch," were his last words. We never heard from any of them again.

I've since been advised by knowledgeable people that it may have been for the best. I was told that if we had technology that the military felt was strategically valuable, they could classify it, leaving us unable to apply it to commercial problems for an undetermined time. This is something to keep in mind. If you think that you have something that might be of military consequence, investigate the potential for military classification before you assemble investors who are counting on a timely return on their investment. Don't be discouraged, though. There are excellent grant programs for your technology if it looks interesting to the military and most technologies don't get classified.

Cultural barriers abound when outside engineers bring new ideas to large well-established companies—commonly referred to as the not invented here syndrome, or NIH. A new technology can be so threatening that not only competitors but also sometimes potential partners will fall prey to inertia and close-mindedness, or even deliberately undermine you.

Early in our history at PAX, I built and tested a number of concept prototypes showing the impact of spiral geometries on fans, pumps, propellers, and drag reduction on fuselage such as boat hulls. I was particularly excited about improving the energy efficiency of refrigeration and air-conditioning, which use around 30 percent of global electrical energy. I'd developed a concept for a novel, biomimetic refrigeration cycle and was searching for an industry partner to codevelop it. My friend and mentor, sustainability expert Amory Lovins of the Rocky Mountain Institute, introduced me to a vice president of engineering at Carrier.

Carrier's original founder was the inventor of air-conditioning in 1902, and the company has become the world leader in its manufacture. An introductory conversation went well, and I was invited to visit Carrier to make a presentation to a group of engineers, including a senior scientist from United Technologies, the owner of Carrier.

Fellow entrepreneurs will confirm that an opportunity to present at that level fills one with excitement and anticipation. I packed my clothes, gathered my prototypes and supporting data, and flew to Carrier's business campus in New York. I arrived a little early and was shown into a conference room. One by one, Carrier engineers filed in—eight in all. The scientist from United Technologies was running late, so we got started and I commenced my spiel with enthusiasm. I explained the background to my work and gave examples of nature's dominance in efficient design. I showed my castings of whirlpools, nature's most streamlined shape, and with hard-to-contain excitement, started to describe my refrigeration concept. The group listened and several asked interested questions. All is going well, I thought. When the United Technologies representative finally arrived, more than an hour late, he asked me to summarize my presentation so far—in five minutes. I was a little dismayed, but I started to do as he requested while the rest of the engineers fell silent. As I talked, he picked up two of my fan prototypes and twirled them in his hands. Like all nature's shapes, PAX prototypes look organic—some people say their curves look a little like seashells or flowers from the garden.

I was beginning to explain my refrigeration invention when the United Technologies scientist interrupted and exclaimed, "This is a complete waste of time. You can't bring me a seashell on a stick (the stick referring to the center axle of my fan prototypes) and expect me to take you seriously. There's no bird that can fly like a Boeing seven forty-seven. Science shits on nature."

I attempted to respond, but all momentum was lost. The meeting wrapped up, and I was shown out of the building. The executive who was my contact continued to show interest, but he retired soon thereafter, so any leverage he could have applied to overcome the not invented here syndrome was lost. I learned from this experience that you can lead a horse to water . . . Carrier wasn't interested, so I moved on. As the industry leader, Carrier wasn't as motivated to try something new as a smaller, hungrier, competitor would have been. The PAX fan that was dismissed as a shell on a stick turned out to be 23 percent more energy efficient and 40 percent quieter than air-conditioning fans used by Carrier. The fully optimized combination of our biomimetic fans and air-conditioning technology has the potential to significantly reduce the world's total electrical energy bill.

Unfortunately, the not invented here syndrome can continue even after a company signs a deal with you and go as far as outright obstruction. Delphi was the world's largest producer of parts for the automobile industry. As competition grew and profits shrank, Delphi looked for ways to diversify, particularly in their specialty area of thermal management. One visionary vice president, Ray Johnson, saw the potential of PAX fans to give Delphi a strong advantage in both computer cooling and air-conditioning. Ray became our champion. So impressed was he with our prototype performance and research that he encouraged Delphi to sign a multimillion-dollar license agreement with PaxFan. We were all excited, enthusiastic, and ready to create great products together. However, our enthusiasm didn't translate to smooth engineering processes. Ray asked his engineers to provide us with specifications for new products. We can only guess that this was seen as an imposition by some on his team. The fan specification we first received was for performance that was completely unrealistic—virtually defying the laws of physics. Ray was told by engineering staff that Delphi had a fan with that level of performance and expected us to improve on it. After a considerable waste of time trying to get clarification, a subsequent visit to Delphi revealed that no such product existed.

Similar experiences occurred over a couple of years. We designed fans and tested them under the conditions that were specified in our engineering Statements of Work. The prototypes would perform beautifully in our facility and be shipped off for testing at one of Delphi's suppliers—who were also often our direct competitors. Unfortunately, on a number of occasions, those tests were conducted under different performance conditions than the specifications we'd been given. Of course, the test results differed from what we'd presented. This cost Delphi and PAX more time, money, and morale.

PaxFan's licensing model was further challenged when Delphi filed for bankruptcy not long into our relationship. As Kenny Ausubel of Bioneers asks with irony, "How do you schedule that into your business plan?" We continued working together, but Delphi's management was understandably distracted and focus on new markets was strained. Two Delphi engineers were assigned to market PAX's computer cooling fans, though for some reason they wouldn't allow PAX engineers to speak directly to their engineers who were manufacturing our designs. This resulted in delays and miscommunications that would have been comical if they weren't so frustrating. Ray Johnson did his best to create new opportunities for Delphi, but in the end was unable to break through the NIH brick wall and Delphi's communications practices. Eventually, PaxFan canceled Delphi's fan license.

If you want to reduce the impact of NIH, you must first cross the divide with professional performance results and good business materials. Beyond that, act early when you hit resistance. Skepticism is not surprising, but if one of your team senses obstruction, even from a junior engineer, immediately call it to the attention of a senior staff member. We could have done this more aggressively with Delphi. Clearly define roles and responsibilities when working with partner companies—and insist on short lines of communication. And if you're signing a commercial agreement, include consequences if your partner doesn't supply appropriate information to help you achieve your assigned tasks.

Another solid obstacle is simply resistance to change—what PaxFan founder and sustainability expert Paul Hawken refers to as *infectious repetitis*. Many industries repeat what they have done year after year, so that nothing will go wrong. Avoidance of risk is paramount; and investment in capital equipment is costly enough that innovation is not seen as worth the price or risk to change over.

Statoil Corporation of Norway is recognized as the world leader in drilling for oil in extreme conditions—places where it requires the energy of two-thirds of a barrel of oil to produce the other third of a barrel for sale. Norway is also the second-largest net exporter of gas and one of the world's largest exporters of oil, so whether at home or working for other countries, Statoil has a well-deserved reputation in oil technology and processes.

The company's senior executives heard about PAX's biomimetic technologies and the potential benefit to the oil industry at an executive education program organized by UC Berkeley. They were particularly interested in the potential to substantially reduce energy losses by deploying more efficient pumps, mixers, or other devices. We were invited to visit Statoil facilities in Norway in early January 2007. It was a fascinating trip for me, since I grew up in Australia and still think of January as midsummer. As it was not long after the winter solstice, the sun rose at 10:30 a.m. each day, swept very low across the southern horizon for about three hours, and settled into a long slow twilight that ended in dark by 4:00 p.m.

Our hosts were gracious and attentive. We were shown through facilities and given presentations; and we presented at a conference with senior staff that included outside suppliers and contractors. I was delighted with the interest shown by Statoil's clearly competent engineers. The company was interested in biomimicry, had extensive R & D facilities, and our domain experts had identified many potential applications in the petroleum industry for PAX mixing, pumping, and separating solutions. A fit between our firms seemed ideal.

Unfortunately, we ultimately realized that while Statoil and other oil companies are willing to take risks in prospecting for new deposits, they're extremely risk averse when it comes to introducing new equipment to their extraction and refining processes. Any downtime through possible failure means huge financial losses and even potential danger to their workers. Statoil had a culture that is not hampered by the not invented here syndrome, and would be open to our products once we had developed, manufactured, extensively tested, and verified the safety of our equipment through a certification society like Det Norske Veritas, Bureau Veritas, or Lloyds of London. Bringing new technology to such a market could be an exercise similar to getting a propeller on a navy ship—requiring millions of dollars and many years of testing and certification before revenue. In the early days of technology development, you can't avoid some dead ends. Our trip to Norway was still valuable; we made connections that may prove helpful in future, and we learned critical information about the industry.

TECHNOLOGY—PLAN FOR PATIENCE

Inventors and entrepreneurs are often so close to their visions that they can be unrealistically optimistic when it comes to projected time and costs to get to profitability. It's a commonly heard phrase that whatever an entrepreneur estimates, one should multiply both time and cost by three. And then add some more.

It's tempting to project rapid uptake of your technology in your business plans. After all, why wouldn't a superior solution be adopted quickly? As the founder of multiple technology companies, however, I can confirm that the unexpected will happen, from test equipment that breaks just before a big presentation to key suppliers suddenly going out of business. Adoption of new industrial technology also usually follows a long, slow curve, as opposed to the sharp spikes of some high-tech or software products. When I spoke recently at an international fluid dynamics conference in Japan, my fellow plenary speaker was a senior professor and researcher from Mitsubishi. He has been studying engineering innovation and particularly the cycle of adoption, from early experiments to broad acceptance. His conclusion is that even in our age of rapid change, a paradigm shift in engineering can take about twenty years to gain full effect. Biomimicry as a discipline is about fifteen years old, and its growth seems to be following the path that the professor outlined.

The takeaway is that it's critical to have contingency plans, both in time and money. Underpromise, so you can overdeliver. On the positive side, a lot of hard work to gain credibility has already been done in this emerging field and newcomers will find the path to success has fewer and fewer obstacles. The future is bright for biomimicry firms that stay the course.

TEAM—CROSS THE DIVIDE

A repeating theme expressed by Biomimicry 3.8's Dayna Baumeister, Calera founder Brent Constantz, and the rest of our colleagues is that biomimics must collaborate in novel ways in order to be successful. While a corporation is considered a legal body, there is actually no such thing as a business. There are only people, working together to create, grow, and maintain a shared vision. Any company is only as good as its people.

There are many approaches to team building, management, and reward. Many start-up businesses hire like-minded friends who share the founder's passion but may not have the right skill set or experience in the target market. This reduces the divide between founder and staff, but doesn't help the company's mission overall. A manager in a larger company that's taking on a bio-inspired project, on the other hand, may be presented with a team with a great range of skills but not a shared vision. At the far end of the spectrum, some businesses conduct extensive, and expensive, recruitment to hire topflight personnel from their target industry, but the difference in cultures from a big, bureaucratic firm to a small start-up can be so great that the new hire gets what's referred to as altitude sickness.

I have a very simple, and fundamentally biomimetic, starting point with my staff and our approach to organizational development. Treat every member of the team, regardless of his or her position, as a unique and important contributor to the total system—not as a commodity. There are many organizations that seem not to grasp this, in my view, critical point. If staff is treated as merely cogs in a machine, instead of valued members of an ecosystem that relies on every part functioning well, there is no possibility that the employer will function optimally—by any metric.

At PAX, we're not joking when we say that we only employ people that we could tolerate being stuck with in a failed elevator for hours. Since we spend as much or more time with our colleagues than we do with family members, and the pressures of a start-up will push everyone's buttons sooner or later, it's important that you find someone far more than tolerable at the outset. We don't require consensus across a department before making new hires, but we do listen closely to any staff member, from administrative assistant to VP, who is turned off by a job applicant. Even the way an applicant relates to our secretary helping with flight reservations for their job interview can reveal a great deal.

If all its internal systems are functioning at their peak, a wild animal survives. If not—it gets eaten. This is a seriously effective feedback loop. A biomimetic company that is entering traditional markets must also have excellent feedback loops. At least once a year, we hold an all-company workshop to improve our communication skills. We also use organizational development tools, such as the DISC behavioral model, which is a survey that reveals individual preferences in communication styles. In combination these practices have been a very worthwhile investment, including reducing information gaps and potential tension between the conservative orientation of many PhD scientists and the more extroverted, persuasive style of sales staff.

TEAM—TURN THE SHIP

People running existing businesses want to expand and strengthen their companies' successes. As we've experienced, they're often willing to hear about new opportunities. Biomimicry lays the groundwork for future profitability and by providing solutions that don't create new problems; it offers something that short-term, cost-saving solutions can't. This can be of real interest to potential clients and partners—particularly during tough economic times. However, when money is tight, most companies consolidate to more familiar ways of doing business. At the other end of the spectrum, when economies boom and order books are full, some managers have less interest in new ideas. It can be especially challenging during such periods for innovative technologies to make headway. For these and the other reasons described, industrial inertia can feel as solid an obstacle as rock.

Ancient Romans found that even the largest slabs of granite or marble can be split with just water and pieces of wood. If you can find a crack and ram in the wood tightly, adding water will swell the wood and eventually break the stone. Overcoming resistance isn't about toppling existing ways of doing business; the trick is to find the cracks in the wall and your team is a key means of doing that.

Beyond its own staff, a company's team includes advisers and business partners. Truly innovative and curious people can be found in every field. Use your networks to identify them and build a technical advisory board for your project. This can be done inside a large organization as well as in a start-up business, expanding credibility without expending extensive resources. Your advisers and their connections will reveal further cracks in the wall of your target market.

Our domain expert strategy has provided PAX with critical insights into potential markets. By contracting with highly respected established professionals in each field that we're contemplating entering, we reduce inertia by educating ourselves in the metrics and value chain of the market. It has paid off in a number of ways. Our industry experts have also introduced us to their networks, with their credibility assisting our entry into successful business relationships.

One way to open doors with potential partners is to have your legal counsel contact their legal departments on an informal basis. Most corporate lawyers are there to save their companies from problems; they're watchdogs who don't bring in new revenue as much as aim to prevent losses. If you can show the legal department of a large company that your intellectual property is secure and that you're willing to execute nondisclosure documents before meeting their team, you can save a great deal of time. The lawyers can then add value to their companies by introducing you to management. This is not the only way to find that crack in the wall to develop a relationship with a large company, but it has worked for us on a number of occasions.

Some men and women who have reached executive status in large companies have real vision and become champions of new ideas. Identifying a thought leader in a client company is a great help, but it's important to remember that these innovative businesspeople are working in a complex ecosystem of their own. They face internal pressures to focus on business as usual and next quarter's profits, as well as having to fight the inertia in their own departments. Make their jobs easier by respecting that they have different priorities competing for their attention. They're sticking their necks out on your behalf, so provide them with prompt and professional information when they need it, and don't be shy about thanking them for their support.

TEAM—PLAN FOR PATIENCE

How often has some relationship turned sour and you thought, I knew I shouldn't have trusted that person? Nature has provided us with exquisitely capable powers of intuition—if we listen. After a career of hard lessons, I have come to the conclusion that I am not prepared to employ someone or do business with any person I don't feel good about. *A smelly fish never improves.* It can be a subtle feeling and it doesn't have to make logical sense. If it starts iffy, I have found that it invariably turns out poorly, and can take more time and energy to get out of the situation than any gain that was anticipated when getting into it. In the same vein, it's amazing how an entrepreneur, who might very carefully conduct due diligence and check the references of a potential employee, will abandon caution and accept money from an investor with very risky strings attached. Planning for patience does mean building contingency into your budgets, but funders—and certainly major funders—need to be considered as cautiously as potential spouses.

I've also learned the hard way not to tolerate behaviors that let down the team. If someone proves to be disrespectful of others and does not correct it, or doesn't have the right set of skills and isn't learning quickly, I let him or her go as soon as possible. It's easy to let things slide, particularly if you're working with friends, but a biomimicry company has enough challenges—it's remarkable how quickly one poor player can pull down morale and frustrate work schedules.

In today's corporate world, people leave or are moved to new positions more often than in the past. Given that a biomimetic project is not a quick fix, it's important to cultivate relationships across your client and partner companies, so that the departure of one key contact doesn't derail your progress. Internally, no matter how small your team, no one person should hold all the details of a project or a process. This is not a reflection on anyone's competence; just the opposite. A bio-inspired business needs to be prepared for a long haul; if one bird leaves the flight formation, the flock must be able to keep on going.

MARKET—CROSS THE DIVIDE
Educating Your Market

Can you sell energy efficiency to manufacturers and consumers who have been saturated with "first-cost" thinking, even if in many cases it's false economy? For example, let's say you purchase a cheap fan, swimming pool pump, or air conditioner that lasts for ten years. When you calculate the cost of electricity to run your equipment during its lifetime, the purchase price can be as little as 10 percent of the total bill. If you purchase a slightly more expensive, energy-efficient device, the overall cost of ownership during the ten years drops dramatically, while the better-built unit usually lasts longer and requires less maintenance. Selling this wisdom is successful to varying degrees, particularly depending on which country you're targeting. South Korean and Japanese customers, for example, will tend to buy a more expensive product that has better value over time. Australian consumers, on the other hand, are known for often making decisions based on the cheapest sticker price.

Many countries are now introducing programs to educate and encourage consumers about their choices. As its website explains, "Energy Star is a joint program of the U.S. Environmental Protection Agency and the U.S. Department of Energy helping us all save money and protect the environment through energy efficient products and practices. Results are already adding up. Americans, with the help of Energy Star, saved enough energy in 2010 alone to avoid greenhouse gas emissions equivalent to those created by thirty-three million cars—all while saving nearly $18 billion on their utility bills." Public utility commissions are now advertising their successes, which helps give momentum to any energy-saving, biomimetic technology entering the market.

Can you sell products based on safety? According to a 2005 study by the RAND Corporation, the death toll from diagnosed asbestos-related diseases in the United States alone is projected to reach more than 430,000 by 2030. Yet the hazards of asbestos have been known and written about since the days of the Roman Empire and were scientifically well understood many decades ago. We're all familiar with "big tobacco" companies denying that nicotine is harmful and addictive and secretly removing reference to damning scientific evidence from public documents. In the 1950s, asbestos companies also tried to maintain sales and limit liability by removing and hiding all references to cancer in employees and users before publishing "scientific research" that they sponsored. Independent and government research papers about the public health threat of asbestos proliferated after World War II. Regardless, until the mid-1980s, asbestos was seen as a miracle material and used in building cladding and insulation, water pipes and brake pads, fillers in baby powder and cosmetic face powders, paint and linoleum, coffee pots and toasters, and five thousand other household and industrial products.

By 1990, about half of all roofs in Australia and many other countries had been constructed of sheeting made of friable, corrugated asbestos/cement. *Friable* means able to release dangerous asbestos fibers to the atmosphere. Asbestos roofs were on buildings including schools, hospitals, homes, and factories. Concentrations of asbestos fibers were easily found on school playgrounds. Worst of all, the asbestos commonly used in Australia was the most dangerous of all varieties—blue asbestos. The cost of replacing the roofs would have been billions of dollars.

Asbestos fiber, in its natural state prior to mining, is coarse, encapsulated in rock, and virtually nonhazardous. Taking a cue from nature, in 1990, I was part of the team that developed and marketed an award-winning, inexpensive, biomimetic technology and process that permanently reencapsulated asbestos/cement roofing and made it safe. We worked closely with the Asbestos Diseases Society of Australia and received their endorsement. We were well funded and had a dynamic, talented, marketing team offering the solution to building owners—in particular, the federal and state government departments responsible for most of the country's schools.

The business model was quite simple: appoint nationwide contractors to apply the encapsulating material and ensure quality control. A typical school roof could be made safe for less than 10 percent of the replacement cost. After all, asbestos-cement roofing was an excellent material for the job—if its fibers could be secured. The government's surprising response? We were slammed with negative press. In an attempt to limit liability, public statements were published that there were no hazards from roofing asbestos—not in schools or anywhere else. Unfortunately, for us, this sent powerful messages to building owners that asbestos roofs were safe and no remedial action was needed. Sales of our technology virtually ended and the company had no option but to cease operations. The Asbestos Diseases Society challenged the government's position, relying on a plethora of scientific reports to represent its suffering, diseased, and dying constituents. A protracted public brawl developed and eventually ended with the truth acknowledged. Buildings were indeed producing dangerous, friable asbestos and school children were at the greatest risk.

Today, most Australian schools and public buildings have had their roofs replaced—at a far greater cost to the community than our biomimetic solutions could have achieved. The lesson here is to thoroughly understand, and not underestimate, the barriers to your business before you get too far. You can't sell a solution until your customer believes that there's a problem.

What to Fix First

There's an old saying that the difference between scientists and engineers is that a scientist finds a solution and then seeks a problem to apply it to, while an engineer defines a problem and then looks for a solution. While oversimplified, there's some truth to this, and biomimetic solutions can fall into either category. However, biomimetic technologies are, most often, solutions that could apply to a number of products, if not industries, so choosing the problem to solve is a defining decision. For example, Sharklet Technologies and BioSignal both developed surface treatments that dramatically reduce the growth of bacteria. The markets they could enter range from antifouling paints to urinary catheters. Our team at PAX understands how to design far more efficient rotors—which means any equipment that has rotating blades like propellers, pumps, fans, mixers, or turbines—and there are hundreds of subsectors in each of those categories. This makes them platform technologies that require time to define, develop, and mature. Determining which market will be most open to accepting and rewarding a new product is critical. In the early days of PAX, we created a product-market matrix to prioritize the impact our technology could have in various markets, correlated with the resources (time and money) needed to penetrate those markets. Our analysis highlighted the areas of fans, mixers, and propellers, so we proceeded to do business development in each area to learn more.

MARKET—TURN THE SHIP

Just like in the natural world, there is a food chain in industry that most consumers never see. Nature and business both want to extract the most value possible with every step. In nature, this results in using the minimum of materials, including by recycling resources. The result in business, however, is a markup in price added by each link in the value chain. The supplier of plastic pellets charges a certain amount to the injection-molding company that makes a fan, who charges a higher amount to the motor manufacturer, who adds its markup to the price on its way to the car radiator, air conditioner, or laptop computer maker, and so on. The players are in a type of economic dance that resists change. This can make the introduction of a biomimetic solution, which sometimes leaps several industrial steps at a time, confusing or threatening to existing players. For example, one of PAX's products can be installed very easily, saving its clients the need to use costly engineering consultants. However, those same consulting engineers are the very ones that the client looks to for advice on what products to buy. That can set up a perverse incentive for an engineering firm to refer their clients to a more costly, less efficient, technology.

Market inertia is not just about internal processes like value chains but also about customer perceptions. For example, the automobile industry has traditionally been reluctant to risk sales and profits on truly disruptive technologies. Their customers like and want the latest gadgets and improvements, but not radically new products. The auto business highway is littered with great technology that was tried but didn't sell as well as expected: rotary engines, DeLorean cars, and the first electric cars, to name a few.

At a macroeconomic level, protectionist government policies, like those that shelter domestic markets from foreign products, can also slow down innovation in outdated industries.

PAX's product-market matrix identified air handling as a good target industry. Fan prototypes could be built fairly inexpensively, international test standards were available, and our competitor's fans could be tested relatively easily against our replacement models. Fan manufacturing is highly commoditized, with a margin of pennies per fan not unusual. But with thousands of different fan applications and billions of fans sold in the world each year, the potential volume and lack of recent innovation in the market made it an attractive market sector.

I found it perplexing when our colleagues and advisers repeatedly suggested that, instead of starting by tackling a challenging problem in air-conditioning or other large fan application, I should start at the bottom and build a new "fart fan" (as one of our creative shareholders called it). The idea was that people dislike the rasping noise and ineffectiveness of bathroom and kitchen range-hood fans, so PAX could design more user-friendly high-efficiency models. I felt more excited by the opportunity when I learned that in some of the cheapest bathroom exhaust fans on the market, only 6 percent of the total power into the system is transferred into usable airflow out of the system. This means that almost all the electricity used to run the fan is wasted. Exhaust fans are also required by law to be fitted to all new dwellings, and about $1.2 billion of domestic fans are sold each year in the United States alone. While humble, that's not a trivial market. Okay, okay. I succumbed to the well-intentioned pressure, and with our clever young engineers, developed a series of bathroom and kitchen fan prototypes that were dramatically quieter, cheaper, and more efficient than the products offered on the market. In the course of our work, I discovered that the U.S. domestic fan market was dominated by one company, which, through its subsidiaries, supplied more than 90 percent of the bathroom and kitchen fans sold to building contractors. We decided to pitch our technology to the company directly, to show how they could supply far more energy-efficient, quieter fans without raising prices to their customers. We should be welcomed with open arms.

Wrong. It took some time to work our way into the organization.

Although their engineers expressed interest in working with us, management seemed to have other ideas. In particular, the company had a number of subsidiaries and offered a wide range of fan brands and models—from low-end, noisy, $12 bathroom fans, up to their quietest models priced at several hundred dollars. At PAX, we created a biomimetic fan that was cheaper to produce than one of their cheapest units, while as quiet and energy efficient as their top models; that is, we could have replaced almost their entire inventory with one design. This would have been good for their customers and the nation's energy bill. The company and its subsidiaries, however, have a large and historic investment in offering a range of products at different price points. This gives customers the perception of having multiple choices. Selling one fan that met the specifications of many price points wasn't appealing to their financial team. And, as was pointed out to me, the company had nearly $2 billion in annual sales and dominated the nation's distribution channels. So no one could really challenge their market position, even with superior product. A company executive did express interest in licensing one of our fan designs as a replacement for one of their most costly, high-end models. I imagined that being able to build these units for as little cost as their cheapest model, but then sell them at a premium would provide a very profitable return. Unfortunately, annual sales numbers for these units were low and the small royalty we would earn for an exclusive license would have been less than our patent and development costs. The result: high-performance PAX fans, which could save the nation significant energy and subsequent CO_2 emissions while providing quieter ventilation for customers, sat on the shelf unused and the market status quo was maintained.

Speaking of air movement and global warming, the methane-loaded flatulence and burps of cows and pigs in a growing number of farms are now collected as they float upward—and reused as fuel. As a global-warming gas, methane is twenty-one times as potent as carbon dioxide, so its use as an energy resource is a good move on several levels. However, there is still the problem of all that gas entering the atmosphere from free-range animals.

Kangaroos could help. Australia, the island continent with an area the size of the United States, is also the oldest land on earth. Over eons, much of Australia has been under the ocean numerous times, so mountains have eroded away and the land has become flatter and flatter. Its soils have been largely leached of nutrients, but there remains a highly diverse, although fragile, collection of ecosystems.

Because of this fragility, the impact of European forestry and farming techniques that were introduced to Australia over the last two hundred years has been devastating. Nearly 90 percent of Australia's native vegetation has been impacted; according to a United Nations report, 75 percent of rain forests and 90 percent of temperate and mallee woodlands have been cleared. Only 6 percent of Australia's land is arable without irrigation, and overgrazing by introduced sheep, cattle, and even rabbits has contributed to large tracts of land turning to desert.

The kangaroo is one of the world's most recognizable animals. About fifty varieties have evolved to suit Australia's unique environments and they're the ultimate animal for grazing the native Australian wilderness (and thriving) with no negative impact. They don't damage the environment with cloven hoofs, like other, non-native, grazing animals. However, with the introduction of farm pastures and year-round water available in man-made wells and dams, kangaroo populations have burgeoned.

The larger kangaroo species are not endangered and provide the leanest meat of any edible animal—with a fat content of about 2 percent. The meat is free-range and kangaroos don't need vaccinations or antibiotics. It is a healthy source of protein, and has a very high concentration of conjugated linoleic acid (CLA), which is being researched for its anticarcinogenic, and antidiabetic properties. CLA is also associated with reduction in obesity and arthrosclerosis.

One of the most surprising and remarkable benefits of kangaroos is that, uniquely among grazing animals, their gut bacteria don't generate methane gas—but acetate, which is a salt. Australia's 120 million sheep, on the other hand, produce 2.4 gallons (10 liters) of methane per animal per day. The 27 million head of cattle produce an average 113 gallons (250 liters) each per day. In 2006, the United Nations reported that livestock produce more greenhouse gases than the entire global transport industry—equal to about 18 percent of all human-related emissions.

This huge eruption of burps and flatulence ends up in the upper atmosphere and is one of the principal causes of holes in the ozone layer—holes that, by allowing in more ultraviolet light, threaten to turn even more of Australia and other environments into desert.

Changing from sheep production and beef farming to kangaroo production could be biomimicry at its best: using nature to heal nature while helping humans to reduce high cholesterol and cardiovascular disease. There's also the opportunity to understand the kangaroo's process of producing acetate instead of methane and potentially replicating it in cattle, pigs, and sheep. André-Denis Wright and his team at the CSIRO in Australia are already developing vaccines in an attempt to neutralize the methane-causing bacteria in a farm animal's gut.

When entering a big market, it's natural and tempting to want to start at the top and solve complex, high-value problems, or address a lower-value product with high sales volume. Instead, we've found that narrower industries, with well-defined market sectors, allow faster penetration in the early days of a technology. This is completely analogous to nature, because every species occupies a specialized niche. If you prove your value in a niche market, you can stabilize your cash flow while attracting corporate partners and distributors to reach larger markets.

A large kangaroo can leap an astounding forty feet in one bound. Entrepreneurs who see a big vision can be tempted to make big leaps over the incremental steps to get there, but the result is inevitably frustration. As one of our business advisers likes to say, "Crawl, walk, run."

Make, Sell, or Have Made?

Once you've identified your market, choosing which business model to adopt is also critical. While it's tempting to trust your instincts, this is one area where using existing business practices and conducting careful market analysis is absolutely essential in confirming your ideas and making good decisions. A primary suggestion: Wherever possible, choose a business model that gives you access to your end users. They're the ones who will reap the benefits of your technology and, therefore, be more willing to pay for it. For example, selling a much more efficient but slightly more expensive fan to someone who will see the savings in their electricity bill every month, like the owner of a computer data center, is easier than selling that fan to a builder who gets his or her profit by keeping initial costs on a batch of new homes as low as possible.

Somewhat surprisingly, at PAX we've found that whole products, or at least high-value subcomponents, actually move more easily into the industrial food chain than isolated parts and components like fan blades. Making whole products is obviously more expensive than components, requiring additional engineering design, materials, manufacturing, certification, marketing, and consequently more investment before getting to revenue. The actual path to the end customer, however, is often shorter and more in your control. At PaxFan we now mostly partner with manufacturers of high-efficiency motors to build finished fan-motor combinations, rather than trying to license just fan-blade designs. We're also designing whole products, like super efficient, quiet, solar-powered fans.

If your biomimetic solution applies to many products, it can feel like you need to become the size of General Electric in order to reach all your potential markets. Rather than manufacturing in a range of markets, licensing is a common strategy to deploy platform technologies. But licensing means giving up control to one or more middlemen, whose own internal politics and issues will affect your success. We saw this with Delphi, which filed for bankruptcy and lost its momentum.

After a number of experiments, we've found that sales (either direct or through a highly motivated distributor) are usually more effective than licensing designs to a manufacturer—unless you can find an ideal licensing partner or you could never enter the final market anyway. Why? Most companies are focused on their own core technology. Their engineers and sales staff are already familiar with and know how to sell it—and their compensation is usually tied to the performance of their own products, not your technology. There is a significant opportunity when you go your own way and build a product: You solidify your credibility and your success will draw the attention of would-be future licensees—who will subsequently be more likely to work with you on your terms.

Following the disappointment with the domestic fan company, we approached the A. O. Smith Corporation, manufacturer of around half of America's hot water heaters and a leading producer of electric motors.

We had learned that A. O. Smith was a key electric motor supplier to Nortek's subsidiaries. Rather than approach their purchasing department, our attorney contacted A. O. Smith's legal staff, introducing ourselves as an innovative company with solid patents and confidentiality agreements. It proved to be an effective way to demonstrate our credibility and find the right managers to connect with.

Electric motor production is highly commoditized and competitive, with very low profit margins. When you make nearly one million motors a day, cutting costs by one tenth of a cent per unit adds meaningfully to the annual bottom line. A. O. Smith was a progressive company and saw the opportunity to increase their profits by fitting high-efficiency fans to their motors and benefiting from either a reduction in materials for the same performance or increased performance and a higher markup. The executives at A. O. Smith were a pleasure to work with—there wasn't the not invented here syndrome that we had experienced at many other corporations. Their company had an R & D department that worked across its two manufacturing arms to improve products, so there were processes in place to help cross the divide with their staff. A. O. Smith executed a license with PaxFan for several cooling fan applications. At their request, we designed two new refrigerator fans to replace inefficient models fitted to millions of refrigerators a year in the United States. The existing fans were designed in 1961 and no attempt to improve them since had succeeded. If we could deliver more efficiency, refrigerator manufacturers could qualify for an Energy Star—an increasingly important marketing feature. We provided a significant drop in energy consumption when we pulled the old fan blade off the A. O. Smith motor and replaced it with a PAX fan. When A. O. Smith matched our fan rotation speed with a customized motor, which cost no more to build, it resulted in an even greater drop in energy and qualified for an Energy Star. As a bonus, it also lowered noise by 40 percent. Obviously, we had a winner.

A. O. Smith was excited. They built mobile demonstration cabinets with fans and instruments to show comparative performance and invested in a promotional video. Years later, not a single fan had been sold. Why not? A. O. Smith didn't have any staff specifically dedicated to selling PAX fans. They would send off the prototype unit to a potential customer, but no one had the allocated time to walk the client through performance testing. An engineer in Brazil working for an A. O. Smith client might test the fan and see that it performed differently than advertised due to Brazil's varying electrical voltage. His question wouldn't make it back to PAX for answers, the engineer would gain the impression that the fan couldn't surmount the local electrical supply challenge (not true), and the momentum for a sale would be lost. Eventually, the license with A. O. Smith lapsed for lack of sales.

Back to the drawing board. After three years of effort, millions of dollars invested, and the best thinking of a talented team, we had developed numerous biomimetic designs that performed better than the world's best—and had even been paid millions in license fees, but PAX fans hadn't yet reached the market. The business model wasn't working. Delays from translation issues, lengthy value chains, and inertia were stubborn obstacles to market penetration, so PaxFan's CEO, Paul Hawken, decided to focus on what we could control: the design time for PAX fans. Fan design is largely stuck in the past with little innovation in decades. When a system that uses a fan is being designed for manufacture, from a hair dryer to a Boeing 747, the manufacturer studies the required fan performance and pulls a model that gives good enough performance off the shelf or orders it from catalogs for mass-produced fans that may have been designed decades before. To be competitive, PaxFan needed to supply not only superior designs but do so in the manufacturer's time frames. Paul suspended marketing of licenses and concentrated on improving PaxFan's internal design tools. The company has now developed proprietary software that rapidly applies our geometries to generate and test fan designs based on client specifications. In the meantime, even without active business development, PaxFan has recently sold a number of fan designs to manufacturers and is preparing for expansion.

PATENTS

Speaking of proprietary software and patented fan designs, there is a school of thought that since bio-inspired inventions are essentially blessings from nature, they should be freely available to all and no one should be able to patent them. As reasonable as that open source model may sound, there are practical reasons why this approach won't produce real benefits and a sustainable future in the short term. Simply stated, effective development of innovation usually requires significant funding. In our capitalist and even in socialist societies, funding is usually not available without the funder gaining some form of exclusive ownership, at least for a time, to provide a return on investment. If, say, a company needs to spend $50 million to develop, certify, and market a new biomimetic wind turbine, but anyone could copy it free of charge, where is that firm's incentive to invest in the program—which, like any development of a new idea, already carries a considerable risk of failure? Patents often mean the difference between a new invention making it to the market and never seeing the light of day. Most, if not all, venture capital funding would dry up if start-ups didn't own patents or trade secrets, to provide protection against competitors.

The idea of a sovereign state granting a temporary monopoly to support innovation is not new. In Venice, patents were being granted as early as the fifteenth century. By registering an invention with the state, which had to be both new and useful, an inventor received temporary exclusivity to make and sell his invention and infringers could be stopped from copying the idea. This was good for the nation because it helped foster new, commercially valuable technologies as well as to keep them in the region where the inventors felt protected in their businesses. Now, about 185 countries grant patents, and the number of patents granted in a country is closely correlated to overall economic strength. China is an exception. Its willingness to uphold patent law is fairly recent and accounts of infringement and piracy are still prolific, but there are some signs that the country is beginning to respect intellectual property rights.

For those who would like to see unrestricted access to nature's intellectual property, there is consolation. All patents expire twenty-one years from the initial application for protection. By this time, the invention should have been funded and developed; returned appropriate rewards to its backers; and can move, fully developed, into unrestricted public use. The practice of patenting ideas may have drawbacks and be an imperfect system, but it does work and it is a way of rewarding effort and risk. There are justifiable criticisms of certain aspects of patent law, particularly as it pertains to abstract concepts like "business methods," or plants and other life-forms; but without patents, technology development, corporate profit and growth, and global job opportunities would contract dramatically—and probably catastrophically.

MARKET—PLAN FOR PATIENCE

Of the three core aspects of a business—its technology, team, and market—the latter is the most out of your control. Despite all the modern tools of analysis, markets are still stubbornly unpredictable. Other players may be working behind the scenes on competing technology; financial pressures (or disasters) in one sector impact many others; and customers often (some say, always) make decisions based on emotions more than logic. Somewhat like the weather or the ocean to a sailor, you can study conditions and plot your course as best you can, but you can never be complacent. Given this, the old saying that time is money has a lot of merit. Venture capitalists describe bringing a technology to market at just the right time as "threading the needle," and push hard to help their portfolio companies reach their target markets before conditions change.

In the case of market penetration, planning for patience means having contingencies in place both financially and strategically—but it doesn't mean sitting back patiently and waiting for the market to come to you. Your job is to establish value in your technology as rapidly as possible—and in almost every case, that means having paying customers. Study your target market in depth, and not just by reading canned market research reports or reviewing leading companies' Web sites, which often don't tell the full story of the sector. Use domain experts to inform you, read industry periodicals, and attend trade shows and conferences.

There's a temptation with some engineers to add "just one more" feature before they believe a product is ready to sell. Resist that. In another part of your team, there's also a temptation in some sales personnel to say they could sell the product if it had "just one more" feature. Resist that, too. Don't go to market with a product that isn't solid, safe, and certified, but use every creative skill you and your team have to get to customers as rapidly as your technology can handle it. That said, be prepared to pivot—if necessary—and shift your focus to an adjacent market if you are blocked in your primary goal.

The good news for bio-inspired companies is that our target markets are not as fickle as the fast-moving volatility of sectors like the computer industry and telecom. While maddeningly slow to change, the industrial world is less about the latest high-tech innovation than about well-proven and sturdy performance. Biomimicry offers the ultimate in performance—which industry is increasingly recognizing. With persistence and stamina, bio-inspired products will reach their markets.

Priorities

What do you want most from your business? A few years after founding PAX, we began working with Andrew Isaacs of UC Berkeley's Haas School of Business. Drew acted as our domain expert in business models. His extensive understanding of technology innovation—and familiarity with hundreds of business case studies—greatly helped us understand our priorities and define our growth strategy.

One of the fundamental lessons Drew taught us was the following: Picture a triangle, with its three points labeled *Impact*, *Control*, and *Return on Investment* (ROI). On this business triangle, you can usually choose two points to maximize, with an attendant reduction on the third point. For example, a small, privately held business has a lot of control over decisions. The owner of the company decides everything from who gets a bonus to what products to develop in the coming years. The company might do very well, with excellent profits (return on investment). But the third point of the triangle, the company's impact, may be less than if the company chose to expand, perhaps by taking on outside investment so that it could ramp up its sales efforts and add more product lines. In this case, the boss-owner might give up equity control, add investors to the board, or even replace him or herself with a manager who had experience at a nationwide level. Control goes down so impact can go up; profits might dip at first if expenditure increases but get even stronger thanks to the extended sales effort. The trade-off can work.

In most companies, ROI is the highest priority from the outset. The board and executives know this and clearly define their actions around it. Control is used to create the highest profits possible and impact is secondary. In other companies, like many nonprofits, impact is the primary priority, with ROI also calculated on the organization's success at reaching as many people as possible.

Can you identify your top two out of three of these priorities? Do you understand the consequences of your choices? Whether a division of a large corporation or a company started to commercialize a biomimetic patent, define and keep the underlying priorities of your business in mind. This will greatly help you and your team to streamline decision making and plan appropriately. Key to this process is also defining what success for your endeavor would look like. Many folks never arrive at an end goal, because they haven't described it and don't recognize the markers on the way. Regularly reviewing the end goals for your technology, team, and market—such as at quarterly strategy meetings—is a simple way to keep on track. When you and your colleagues share the vision of where you're headed, you're far more likely to get there.

As the examples in this chapter show, the path to business success is never a straight line. After years on the water, the best analogy I've experienced is that of a sailboat on an ocean crossing. Your boat is your technology, and your business will sink if it isn't sound. It doesn't need every last gadget, but it needs to be fully functional and fit enough to handle what nature throws at you. Yet the soundest boats will go nowhere without the captain and crew setting their hands to the wheel, the sails, and the rigging as an interdependent team, relying, with their lives in some cases, on the knowledge that each will handle their positions with skill and zeal. When the desired harbor is defined and the course is charted, then it's a matter of responding to the weather and the seas—like the market that will fill your "sales" when you set your course right, or knock you down if you fall asleep at the wheel. By taking on board the ways in which a biomimetic business differs from other start-ups, I'm confident that your trip can be easier—and a lot more fun.

Chapter 10

DOLLAR SIGNS

Man's best friend is living up to his name in biomimicry. Domesticated at least fifteen thousand years ago in East Asia, dogs have already hugely benefitted their masters' survival chances with their contributions to defense, hunting (finding and flushing out prey), and cleaning up campsites campsites from disease-attracting flies on food scraps and bones. Added to these more basic duties, it has been authenticated that the owners of dogs have, overall, "fewer cardiovascular, behavioral, and psychological indicators of anxiety" than nonowners. Since the eighteenth century, dogs have been used to calm and socialize patients with mental disorders. ADHD and autistic kids do better in the company of dogs, and dogs have been found to have measurable calming effects on prisoners.

Now scientists have discovered that some dogs, with training, can detect human skin cancer through the smell of lesions, and breast, lung, bladder, or bowel cancer by the smell of a patient's breath, urine, or stool samples. Researchers believe this is because cancer cells give off specific gases that dogs' acutely sensitive noses can differentiate. Dogs have also shown the ability to give early warning of wound infection by dangerous streptococcus bacteria. While humans have five million scent receptors in our noses, dogs have twenty-five to fifty times more than that and can detect odors at concentrations one hundred million times lower than we can. Bloodhounds have two hundred and fifty million receptors and can pick up signals from drops of blood that are ten years old. Early successes with canine cancer detectors hold great promise for noninvasive screening tests.

These highly desirable abilities in real-life dogs have inspired a number of biomimetic research programs to develop electronic noses. We're all now used to the swab tests conducted by TSA at airports looking for explosive residue. The Pentagon spent $19 billion over six years on the Joint Improvised Explosive Device Defeat Organization—tasked with finding the best bomb detector. It turned out to be a dog. Israeli authorities have now implemented a screening technology at border crossings, DiagNose's RASCargO, that works in conjunction with dogs, analyzing air samples from vehicles.

For centuries, people have believed that dogs can give early warning of impending earthquakes. The U.S. Geological Survey confirms that dogs can feel a primary shock wave that arrives 60 to 90 seconds before the larger tremors that are felt by humans. Dogs have also proven themselves capable of reliably predicting the onset of epileptic fits in their owners. If this ability can be replicated in a patient-worn device it would give great security and freedoms to sufferers.

The potential benefits and profits from mimicking animal sensory systems are obvious, though I find it ironic that dogs, with their smelling acuity millions of times more effective than ours, invariably sniff each others' butts in the closest proximity possible.

Whether electronic noses, hippo-inspired sunblock, or planes that can fold their wings, phenomenal biomimetic opportunities are emerging in labs around the world. Most of the discoveries described in this book are waiting to be commercialized. Like Paul Stamets or Brent Constantz, sometimes a university researcher will become an entrepreneur. Because not all professors want to become businesspeople, however, most campuses also have departments that coordinate the transfer of technologies to outside companies that want to develop them. Alternatively, published scientific research can inspire creative thinkers to dig further, learn all they can about the topic, and create new products.

Whichever way, bringing a bio-inspired idea from the lab to the market means spending money. It really doesn't matter if it's a biomimetic innovation or in any other field, the rules in the funding game are the same—and to borrow a line from our canine companions, it can sometimes feel like dog eat dog.

I've met many inventors and aspiring entrepreneurs over the past thirty years. More than a few believed that having an idea—or securing the rights to it, if it was someone else's original thought—meant their job was done. The expectation was that they could be paid handsomely for the concept and others would do the rest. I have never seen this approach result in anything worthwhile. They say that success is 1 percent inspiration and 99 percent perspiration; this is 100 percent my experience. Without full knowledge of your subject—or at least an abiding determination to learn—and passion, conviction, and total commitment, you shouldn't give up your day job.

I have also always found a way to fund the earliest stage of a venture myself. The primary stage involves understanding and describing the technology. What is its position and possibility in the marketplace? What are the competitive technologies in the domain? How is the market currently set up? Who would benefit from the technology, and who would pay for it? This phase also requires the preparation of presentation materials with compelling models or concept prototypes. These don't necessarily need to be working models, but must be well-thought-out and demonstrate or describe feasibility. The initial stage of a venture also includes paying for the preparation and lodging of patent applications. With all that in place, the real work begins.

The requirement for further funding can be answered in numerous ways. In my view, mortgaging your house, or your parents' house, is unacceptable. It is highly unlikely that you can accurately determine the future financial needs of your venture at an early stage. Any mortgage is unlikely to cover your costs to profitability and you will fall short. The stress and desperation caused by this to you and your family will severely impact your judgment as you arrive at future decision points. If you can demonstrate that your idea has real merit, it is far better to spread the risk by including third-party funding in your strategy.

BUDDYING UP

One option is to find a corporate partner. This can be an excellent resource, though it's often easier said than done. First, you have to find a company that is open to outside ideas. In the cases where we attempted to connect with these corporations early in PAX's history, we discovered that although we had a valuable technology key, one that could directly improve their products and bottom line, there was no lock for it to fit. Contrary to some companies' brand advertising, many large corporations in America have little remaining research and development capability. A lot of "innovation" focuses on how to improve the cost efficiency of distribution channels or trim prices for raw materials.

On top of that, PAX technology was such a radical departure from conventional thinking that even the most well-intentioned manufacturing engineers did not understand how to adapt their processes. Instead, with a flood of cheap imports from Asia and Mexico, U.S. mechanical engineering staff were frantically working long hours and shaving costs just to keep their jobs. They had little or no bandwidth to try and understand, absorb, and develop new ideas. This can be a real challenge for biomimetic products. As I was told by an exasperated chief engineer, "How can I be expected to work a sixty-hour week to keep my job from being outsourced and on top of that be expected to implement a new technology that I don't understand?"

Over and over, I was advised that the only way big companies could work with us was if we were suppliers of finished product. But that meant we had to become manufacturers—with manufacturer's expertise, facilities, finance, and management—in every field we had explored: turbines, refrigeration, pumps, fans, propellers, fuselage design, and more. Although daunting, we saw no alternative but to head in that direction. Our business model was evolving as an intellectual property company that would launch and fund subsidiaries or licensee partners, each of which would specialize in particular industry sectors. Once we had products on the shelf, we could expand our businesses through relationships with corporate partners. This has worked well for PAX's subsidiaries and is a good example of planning for patience. Don't expect companies to be able to help in ways that they're not set up to do.

ON THE WINGS OF ANGELS

If not corporate funding, where else can a start-up business find seed capital? PAX Scientific was founded as a common stock company. It was eligible to sell stock to investors without complex securities filings, if they met certain minimum net worth requirements. A series of fellow entrepreneurs, accountants, lawyers, doctors, engineers, and other successful businesspeople joined us as shareholders in PAX and have truly been our angels. While we could offer no promises for when they would see return on their investment, none of us expected that PAX would take more than five years to reach profitability. Their support and understanding of the challenges we've faced exemplifies the term *patient capital*.

Entrepreneurs often source money from angel investors in the early stages. I am a great supporter of this strategy. However, while the amount of paperwork may be less than with an institutional investment, there are clear and precise regulations regarding this kind of fund-raising, and it is essential that they're complied with. California and most other states want to prevent aggressive entrepreneurs from taking investment from little old ladies who don't have much money or from enthusiastic but cash-strapped working people. Investors need to be qualified by having enough assets or annual income to survive just fine should the high-risk venture fail. No one should invest dollars that they can't afford to lose. They must be fully informed that the investment has significant risk and might never be returned. I have seen some entrepreneurs cut corners here. This is foolhardy and not the way to build an ethical, sustainable business. There is no shortage of great sayings about this subject, including, "As you sow, so shall you reap." It's also been said, "Honor is a gift you give yourself." Never compromise on ethics, no matter what others seem to be doing or urge you to do. It's simply not worth it.

Angels offer other benefits than being a source of funding without extensive paperwork. Often they invest because they believe in you personally, your business vision, and your mission. I look for angel investors who can contribute not only dollars but also experience or connections they're willing to share. Having your funders assist in such a way is a benefit that is hard to overestimate. Of course, as with all businesses, there can be a fine line with some well-intentioned investors between helping and over involvement in your management. My good friend Iain Stephens is a particularly skilled shipwright. When he quotes a job for a new client, the boat owner will often ask what it would cost if he pitched in to help. As Iain says, only partly joking, the price goes up by 30 percent—for the aggravation. However, if you've done your homework and understand your technology, your market, and your priorities and communicate clearly, angels can greatly help without hindering.

Early angels typically invest between $10,000 and $100,000. As the project develops, amounts of $100,000 to $500,000 can become accessible. PAX has had the tremendous good fortune of receiving $1 million from an individual who deeply believed in our mission. However, angel investors of this size are rare. As a consequence, unlike some venture capitalists, angels don't tend to aim for control of a company. Given the reality that bio-inspired businesses need time to mature and define their markets, this form of patient funding can be of critical importance.

NOTHING VENTURED . . .

It is hard to overstate the benefits of the venture capital industry to America and the world in general. Some research suggests that VC-backed company revenue, including from such cybergiants as Amazon, Yahoo, and Google, account for 21 percent of current U.S. gross domestic product. Microsoft and Apple, two of the most valuable companies in the world, were both founded by start-up entrepreneurs who received venture funding.

Venture capital evolved to fund the early stages of high-risk, high-reward opportunities. VC partners are always hunting for the next Facebook, or its equivalent, in sectors including software; hardware; telecom; health; biotech; or, increasingly, clean tech.

There are around two million new businesses started each year in the United States. Fewer than one thousand receive VC funding (a chance of one in two thousand). Typically, fewer than one hundred of those portfolio companies will create really significant wealth (one in ten VC investments). These are steep odds, so venture capitalists have developed fierce survival strategies about how, and in what, they invest their funds. They seek as high a return on their investment as possible in as short a time as possible—hundreds of times their investment within three years, if they can get it. Remember that a VC firm typically sees one significant success out of ten start-up companies. Your start-up company, if successful, must therefore make enough profit, and the VC must have enough ownership, to compensate the investment firm for their other nine failures. That puts a lot of pressure on you to deliver.

In the heady early days of the Internet in the year 2000, there were 1,022 active VC firms in the United States. These numbers have more than halved in the past decade, to 462 active firms in 2010. Staff was slashed in the wake of the worldwide financial crisis; just in the first quarter of 2010 the amount of funds raised dropped by 49 percent compared to the year before. This severe contraction was in step with the larger economic recession. Which one causes the other is hard to determine, however, since no investment eventually equals no growth. The VC industry is rapidly growing in China, India, and Brazil; and their economies have been booming.

It's the dream of many aspiring entrepreneurs to be selected by a VC firm for investment. VC interview schedules are full. Although consistently tough in their negotiations and expectations, there are VCs with solid reputations for fairness, genuinely respecting the needs and welfare of their entrepreneurs and staff. On the occasion that investment is received, it can be a great boost to an entrepreneur's dream—or sometimes turn into a nightmare. There are, unfortunately, VC partners and firms whose predatory behavior has earned venture capitalists the nickname vulture capitalists; though, as a naturalist, I know that vultures at least wait until their dinner is dead before dining.

Due to the industry's varied reputation, we'd elected, early on, not to include VCs in our fund-raising models. For example, a VC's form of stock usually comes with the right to approve or veto new investment terms and therefore has a tremendous amount of power at future fundraisings. By 2007, however, PAX had so many technologies evolving from nature's streamlining principle that we wanted to accelerate our path to industrial markets. After a year of planning and negotiations, we spun out a new company called PAX Streamline and contributed a number of our patents to the company. The original PAX shareholders retained majority ownership of PAX Streamline after allocations for the staff and our VC investor.

Without going into a detailed history, suffice it to say that over the next four years, a series of corporate gyrations spanned five CEOs, cycles of rapidly accelerated spending followed by layoffs, more than $1 million in lawyers' fees, the involuntary reduction in our ownership of the company to 10 percent and the increase in the venture capital firm's ownership to 70 percent. In the end, conventional engineering was pushed, not a single biomimetic technology that had been contributed to the VC-backed company was developed, and the engineering team couldn't produce profits on a venture capital timeline. After more than $25 million was spent, the company was closed at a complete loss to all stakeholders. Would I or any of my team choose to do this type of transaction again? Not a chance in the world. Regretfully, I've met other entrepreneurs who've had similar experiences.

Some VC firms focus only on businesses that could earn billions of dollars in revenue. A company that grows steadily, but reveals less potential value than originally hoped, gets drowned like an unwanted kitten—which is another insider phrase in the VC world. For example, a VC-backed rapidly growing California company was building patented products, with initial sales to satisfied customers and millions of dollars in orders lined up for the coming year. An experienced CEO and skilled team of engineers and marketers were working well together. The company's products were ideally suited for U.S. manufacture. The CEO now needed to raise $10 million to get to the stage of profitability. Sales projections were estimated at $300 million per year within five years, and a number of investors were willing to commit funds to support growth. A nearby county had also offered millions in incentives if pilot manufacturing was centered there in order to encourage local employment. However, when it came time to strike a deal with the California manufacturing site, the board members representing the VC firm that had funded the project up to this point were categorically opposed to manufacturing in the state, insisting that it should be located in a country such as Vietnam or Mexico, and would not vote to approve the plan.

As the CEO recovered from that setback and continued to seek expansion funding, the primary VC announced that the start-up's $300-million-per-year projected revenue was too small for their firm's investment portfolio. After weeks of high-tension negotiations, with management and staff continuing to work but uncertain of their future, the VC firm chose not to invest further funds themselves and also used its veto rights to refuse investment offers from other VCs who were interested in a $300 million market. Why a VC firm would do that is hard to understand, since it meant that no one would get any financial returns from the technology, even if smaller than hoped, but the VC had that power and there was nothing the company could do. The young business ran out of money and was forced to cease operations, laying off all staff and leaving creditors unpaid.

Don't get me wrong; it can be a great benefit to have clear, exacting, and unemotional input into your company's growth strategy along with funding. This is a powerful engine for the capitalism that built England, the United States, and the developed world. However, there are some VCs that promote themselves as an inventor's saving grace. They promise to provide management expertise, ranging from finance to marketing, and introduce you to business connections that will greatly increase your chances of success. But it is not a guarantee of a successful outcome. Bear in mind that many venture capital partners are very busy, with many portfolio companies to supervise. Most have no scientific expertise in a start-up's technology, and may not even be familiar with the intricacies of its target market. They do know the importance of building a good team, however, and to focus hard on the critical path to market and consequent revenue.

Some venture capitalists urge their portfolio companies to brainstorm how they could accelerate development—even ten times faster. "Get the best people and move as quickly as possible. Don't worry about the cost; we'll help you out if you run out of money. We're in this together." I've heard this on numerous occasions. I fully support getting the best people possible. As for speed? It can be likened to keeping nine pregnant women in a room for one month and expecting a baby. This may work in certain cases, such as when millions of lines of software code need to be checked for bugs as quickly as possible; and since venture funds are set up to return profits to their investors in just a few years, their aggressive timelines are understandable. But they don't always suit the development cycle of bio-inspired products. Medical applications; industrial equipment; and solar, wind, and other renewable technologies can require months or even years of trials to demonstrate their performance against traditional solutions. In a push to get to market and increase the value of the company, engineers can be rushed into short-term decisions, cutting corners, and making assumptions that inevitably slow down high-quality product development.

Unfortunately, when funds inevitably run low due to these ramp-ups in expenditure, the now close-to-insolvent company must hustle to raise more capital. The urgency to raise funds makes it a buyer's market, and the company's valuation, or price at which new investors can buy stock, slides downward. The next step—the original VC "rescues" the firm by putting in more capital—but in the process, it devalues the company and ends up with most of the equity. Unfortunately, I've lived through this myself.

Many VCs are networked with one another, and one venture firm doesn't encroach on another's turf. I've heard VCs say that they won't offer to invest in a new round of funding into a company unless any earlier VC investors informally agree to the terms behind the scenes. Meanwhile, early angel investors, the inventor, and the employees are often holders of common stock, which comes with less powerful rights. They have no way to control the negotiations that are possible between VCs and are surprised and disappointed when the financing terms offered dramatically reduce their portion of equity. Hence the term *vulture capital*. If you've seen the movie *The Social Network* about the Facebook story, you've seen this scenario played out.

Having said all that, there are VC firms with ethical reputations, run by sincere personnel. They know their top priority—return on investment—and they're tough but fair. How do you weed out the vultures from the worthwhile partners? Don't be starry-eyed. It's essential to do exhaustive due diligence and consult with specialized legal counsel before committing your technology to a VC contract. Interview management and staff at other companies in the VC firm's portfolio including some that failed. Research the history of how employees and other common stock holders fared as the companies grew. When in doubt, listen to your gut and speak up—and get any promises in writing.

As noted earlier, the time for significant return on investment is usually proving to be longer for clean tech and bio-inspired products than for software. Google and Facebook didn't have to get permits to build a green city, rent cranes to install innovative solar panels, or pass lengthy clinical trials for nontoxic pharmaceuticals or medical adhesives. A venture partner should demonstrate not just lip service but a real understanding of the timelines that are realistic to penetrate your target market, and be willing to put it in writing. If your business plan has true potential and you do careful homework, you may find a venture partner who will strike a fair deal. Otherwise, the price you may pay is not worth the money.

A HELPING HAND

The U.S. governments, both federal and state, offer thousands of grants worth billions of dollars each year. Virtually every department has funds allocated to grants, and many other countries have similar programs. Grants are also offered by many other entities, ranging from utility companies to nonprofit organizations.

After decades of watching this new field of biomimicry from its inception, I am convinced that grants are an essential way to promote its growth. If you qualify and play by the rules, grants can be a wonderful—albeit usually slow—source of nondilutive risk capital. PAX has had very positive experiences with most of its grants, totaling millions in funding over the past five years. They have enabled us to create tools that are now meeting our customers' product needs much more quickly. This translates to economic benefit for PAX, its staff and clients, and the country. A fringe benefit of receiving grants is credibility. It demonstrates that you can withstand in-depth due diligence, stick to budgets, and provide disciplined and comprehensive reports. Potential business associates and clients see success in this area as an endorsement of your professionalism. Apart from anything else, the morale boost you get from receiving a grant is priceless.

Unfortunately, finding a grant that you believe your venture qualifies for and then preparing a successful application is not usually a quick or simple process. While the granting body doesn't intend it that way, it can sometimes cost almost as much in time and money to apply for a grant as the benefits received. This is in large part because governments are severely criticized if grant programs are seen to be squandered or exploited by less-than-scrupulous recipients. You might remember the scandals some years ago involving General Dynamics and their overbilling of hundreds of millions of dollars on various programs. The government was charged $700 for a hammer and billed for an executive's use of a dog kennel for his pet. In fact, in 1985, nine of the U.S. military's top ten suppliers were under criminal investigation for fraud against the government.

Now, as you can imagine, there is a whole lot of red tape that needs to be dealt with before a scandal-averse bureaucracy is willing to write you a check, and there is no shortage of public and political criticism if they get it wrong. The failed California solar company Solyndra is a case in point. The Department of Energy (DoE) underwrote a $535 million loan for its growth—one of a number of loans to the emerging green economy. Unfortunately, the company went broke, in large part due to changes in the price of a competing raw material and the global economic downturn. It became a political flash point with the opposition blaming the Obama administration for the failure. However, out of $40 billion deployed to date in the DoE's loan program, the Solyndra losses represent less than 1.5 percent of the overall funding. That's a lower default rate than U.S. mortgage, credit card, or student loan lenders experience. In the same time period, Chinese government loans to solar companies alone totaled nearly $30 billion—fifty-five times the Solyndra investment, and three times the total of DOE loans to U.S. solar companies.

Should you be selected as a government grant recipient, budget slow-downs can still delay the actual release of funds—sometimes to the point that the research you planned to conduct is no longer needed. This happened when PaxFan was awarded a $300,000 grant by the California Energy Commission to design more efficient fans for computer servers. We identified a California-based global server manufacturer as our partner for the grant and lined up our staff to start work—just waiting for the go-ahead from the grant committee. Eighteen months later, the funds were finally okayed for release. By that time, our partner had already finalized their server design, without biomimicry, and released its new, nonoptimized product to the market. We had to scramble to identify a new partner and consequently change some of the grant tasks. The funds, and our research, were held up another six months while we waited for the Energy Commission's grant committee to approve our requested changes. During that waiting period, we applied to the same program for a grant to work on a different project. Although well qualified, we didn't receive it because we already "had" a grant, though twenty-seven months after being awarded the grant, we still hadn't received any funds. These kinds of delays could cause a small business to fail. Do not spend grant money until you know the program is funded and operational. Then test the system to be sure you understand the intricacies of its reimbursement and contract rules.

Despite the occasional frustration, grant funding is a valuable tool for a bio-inspired company and can seriously help launch our new economy. I hear repeatedly from my colleagues that their greatest need is not expansion money to ramp up manufacturing or distribution, which can be gained from banks like any other company, but seed capital to get from the scientific research stage to tested production prototypes. Grants allow a company time to cross the divide, turn the ship, and plan for patience—the right recipe for long-term success.

The U.S. Department of Defense budget for 2010 was nearly $700 billion. Sustainability experts, Dr. David Orr and Amory Lovins, are now advising the U.S. military on national security—including how to learn from nature to ensure a safe water and food supply, as well as a decentralized energy grid. If the United States invested the equivalent of just a tenth of 1 percent of its annual defense budget ($700 million) on seed grants for biomimicry, it would result in hundreds of new products that would transform industry and the nation's dependence on foreign oil. If the United States doesn't fund biomimicry research, other countries will—and, actually, they already are.

THE ROAD LESS TRAVELED

Any entrepreneur will tell you that bringing an innovative product to market involves a great deal of dedication, long hours, and endless problem solving. New ideas take a whole lot of effort to manifest. There are extended periods where all you seem to do is put out bushfires with little forward progress. On top of seemingly endless surprises to your carefully laid business plans, there is the recurring specter of running out of money, which will earn you the loss of your dreams, disappointment of your staff, and criticism from your investors. How do you fortify yourself to handle the stress? In my experience, echoed by innovators whom I've worked with over the years, there are two actions that separate a successful and satisfying business (and thereby a more satisfying life) from the other options available.

First, find something you truly believe in and love doing that you would pursue even if you weren't being paid. Otherwise you'll burn out, resent it, and shortchange yourself, your family, the project, and all the supporters you have enlisted. If you're not fascinated by your project, you may be working on it out of fear that it's the best you can find, or out of greed—whether for money or status. On the other hand, the pursuit of something that you enjoy, which has benefits for others as well, will sustain you when the going gets tough. You're far more likely to create excellence and be seen by others to be sincere and dedicated. It's easier to find support from investors, colleagues, and clients when they see genuine passion in you.

The second and perhaps most important key that I've learned to unlock fulfillment is a decision. First, take the time to imagine and list the essential qualities you want in your life, things like integrity, honor, fun (and the time for it), and your standard for compassion and tolerance. Make the decision never to compromise on these—no matter what. Then, roll up your sleeves and get to work.

In the fifteen years since starting PAX Scientific, we've learned that overhauling the industrial world is complex. A wise man once said, "It takes twenty years to become an overnight success." Another said, "The difference between success and failure is perseverance." The examples in this and other chapters aren't about easy victories but reflect a slow yet steady shift to a new paradigm. PAX and our biomimetic colleagues have been blazing trails, stepping on rakes, and—yet another start-up expression—crossing the valley of death. The huge wealth and value-creating potential for biomimicry is worth the effort. As the problems created by climate change and energy and resources deficits only become more severe and big business sees that business as usual is no longer effective, the path for future bio-inspired businesses will be easier.

I wish you a deeply rewarding journey.

Chapter 11

REORGANIZING (YOUR) BUSINESS

WHAT YOU AND A REDWOOD FOREST SHOULD HAVE IN COMMON

When Americans refer to filing Chapter 11, they mean that a financially troubled business has filed documents asking the courts for permission to reorganize. A bankruptcy judge can grant a company a temporary reprieve from its creditors, so that the business can continue to operate while it attempts to work its way out of bankruptcy. Much of our world is in a virtual Chapter 11, with total debt three times that of global gross domestic product. We need to reorganize—to reinvent our businesses and our economies in the face of increasing energy costs and environmental pressures. This can be achieved through biomimicry, not only in our products but in the way we run our companies.

How about running your business like a redwood forest? That sounds like a tree hugger's mantra, but a growing number of high-performing companies around the world, including those that currently have no plan to sell biomimetic products, are seeing immediate gains from doing exactly that.

A mature forest is a fully self-sustaining producer of diversity and abundance. Many businesses, however, function more like invasive weeds. Their strategy is to spread rapidly into an area, put down shallow roots, and use more than their share of local resources. Invasive weeds are often species with short life cycles. They grow fast, using sunlight, water, and nutrients, and dominating more finely tuned members of the ecosystem. In some cases, they can overwhelm their host habitat.

Without trying to oversimplify, the analogy is easy to see. Our forefathers moved into new territories and ecosystems, expanded quickly, used up resources, and created a lot of trash. Next generations spread farther into fresh lands. Many businesses do the same. It's a common lament that businesses tend to focus on how to survive in the short term, rather than thinking strategically about their long-term prospects. Some companies, particularly in Japan, do have hundred-year business plans, but many Western business strategists don't anticipate beyond two to five years.

Still, a weed's strategy does work in many ways and has worked for humans and businesses, too. Until we started to run out of resources and room. We see this at work in the world's tropical rain forests, where the land is stripped of life, burned, and converted into beef-raising pasture for hamburgers (two hundred thousand acres every day). However, rain forests often grow on poor soils, with nutrients washed away by millennia of incessant rain. The rain forest ecosystem relies on a thin topsoil of decomposing vegetation, which rapidly disappears after clearing. The land soon fails to produce good pasture and is so degraded by erosion that the farmer abandons it and clears the next tract of rain forest.

Quite simply, with so many of us on the planet, we can no longer live like invasive weeds. We've been growing monocrops and business monopolies that are stripping out economic and ecosystem diversity. The good news is that without compromising our quality of life, just by making different choices, we can shift our niche and live more like mature trees in a forest ecosystem. How is that more sustainable? Instead of spreading out shallowly over lots of ground, trees root deep and are large and stable, with long, complex life spans. They're adapted to be a part of a weblike food chain that supports far greater species diversity than a field of weeds. And unlike a field of weeds that can be wiped out if weather conditions turn nasty for more than a short while forest ecosystems can survive and thrive for millennia—in the face of winters, winds, and even drought or fire.

How can a company, or an entire economy, conduct business like a well-established forest? In her book *Biomimicry* Janine Benyus summarizes the research of ecologists, economists, and sociologists (the blend of the three is now called industrial ecology), and describes ten key ways to operate like a forest. Underlying them all is the main goal of life itself: to create conditions conducive to further life.

1. Use waste as a resource

2. Diversify and cooperate with other species to fully use the habitat

3. Gather and use energy efficiently

4. Optimize rather than maximize

5. Use materials sparingly

6. Don't foul your nest

7. Draw up instead of down on resources

8. Remain in balance with the biosphere

9. Run on information

10. Shop locally

Some of these principles may be less pertinent to your business than others, but you can apply many for immediate impact. Of course, your family is a mini-business in itself, so these lessons don't need to be left at the office. At PAX, we addressed one strategy per month by circulating a summary to our team, discussing it briefly at a staff meeting, and letting suggestions come from the group about how to implement each concept. It was painless, productive, and engaged us in creativity. I am continually impressed by the innovative and practical suggestions of staff at all levels.

Most of the topics are self-explanatory. Let's take a brief look at them, though I won't try to re-create Janine's exploration in detail. I urge you to read her book *Biomimicry* for that. It's a great resource for anyone who wants to learn more about this field.

Use Waste as a Resource

What we call waste, nature sees as indispensable building blocks for other processes. Janine comments that, ". . . as a system puts on more biomass, it needs more recycling loops to keep it from collapsing," and goes on to describe how industrial ecologists are attempting to build "no waste" economies. This makes great sense. Waste is usually lost opportunity that can be repurposed to restore balance to a system and produce wealth.

Any business can not only reduce but also recycle and reuse much of the waste that we all unavoidably create. Beyond looking at the nuts and bolts of how your company makes or sells its products, this lesson can also apply to individual choices. This doesn't mean sneering at colleagues who don't compost their lunch leftovers or refusing to use scratch paper unless it's from the recycling bin. As our adviser Paul Hawken told us when we first started PAX, it's laudable to recycle everything from paper to paper clips, but don't lose sight of the big picture.

Waste isn't just about garbage—as the old saying goes, one man's trash is another man's treasure. Are there redundant or underused products that your company can sell or donate to others who need them? Is your company's computer room air-conditioned, even in the winter? At PAX, we were spending thousands of dollars each month to cool the tightly sealed room that holds our supercomputers. We now direct the heat generated by our computers into an unheated lab space in the winter, which increases the comfort of our staff and lowers our electricity bill.

A world-famous complex in Kalundborg, Denmark, has taken waste reduction and modeled it on a living forest. Over the past thirty years, a number of businesses have partnered to evolve very profitable symbioses. Included are Denmark's largest power plant and oil refinery, a factory that makes fifteen million square yards of construction plasterboard a year, a pharmaceutical company that produces 40 percent of the world's supply of insulin, and Kalundborg's municipal heating and water system, which warms twenty-thousand homes. Each company has found ways where the waste from another firm—whether excess heat, fly ash, sludge, or even hot salt water—can be reused. The businesses say that good communication between managers, helped by being located near one another, is the key to functioning partnerships and business contracts that are beneficial for all the parties involved.

Using waste as a resource can support new opportunities for your own or other businesses. For decades, fast-food restaurants paid to have their spent fryer fat dumped in landfill. Now it's the feed stock for biofuels. Old but still fully functional cell phones are being reprogrammed and given to seniors or soldiers on military duty at a greatly reduced cost. Entrepreneurs and big companies alike are now recovering the precious metals built into computers and other high-tech devices.

You can recycle almost anything, just as a forest does. Nicholas Tee Ruiz, a staff member at New York's Museum of Modern Art, has even crafted a business making bow ties out of discarded materials— including Lego blocks, soda cans, and a host of other unlikely waste. A crew from the environmental nonprofit Adventure Ecology crossed the Pacific Ocean in Plastiki, a catamaran made of 12,500 plastic bottles. The EcoARK, commissioned by the Taiwan-based Far Eastern Group, is a three-story, $3-million building featuring an exhibition hall and amphitheater—built out of one and half million recycled plastic bottles. The designers claim that it can handle hurricanes and earthquakes, and is the world's lightest, moveable, breathable, environmental structure.

EcoARK in Taipei, Taiwan

Ugandan refugee Derreck Kayongo collects millions of slightly used bars of soap left in American hotel rooms, melts them down into new bars, and gives them to impoverished people in Africa. With more than two million children dying each year from diarrheal disease, largely preventable by washing hands, Derreck is using wasted soap to save lives.

What waste-reuse partnerships could your business create? Ask your staff for ideas. At PAX there is no such thing as a bad idea from a staff member. We listen to all suggestions—across departments—because the kind of thinking needed to solve a problem is often a different type of thinking from the type that created it. Often the craziest ideas hold clues to creative, viable solutions. Who knows, you may come up with a recycling idea or process to build a runaway success on—it's a big resource out there.

It's also worth knowing where your waste is going when you do recycle. For example, one fifth of American industrial and automotive car batteries are recycled in Mexico for their lead. However, Mexico has lax pollution control regulations, and health workers are now finding that children living near the recycling plants are experiencing chronic lead poisoning. You can choose local, responsible recyclers. Incidentally, the most recycled materials on earth? According to the Federal Highway Administration, it's asphalt, at 80 percent, followed by aluminum cans at 60 percent.

Diversify and Cooperate with Other
Species to Fully Use the Habitat

A Bioneers tagline about nature is that, "It's all relatives." This is to me the clearest and most important lesson we can learn from biomimicry. Everything is interdependent. Everything in nature, including us, is in constant relationship with other organisms (including people) and the habitat itself. Since scientists are increasingly confident that we all came from just one thousand breeding pairs seventy thousand years ago, we're all literally relatives as well.

This principle underscores the survival imperative of cooperation. For example, can you describe your business ecosystem? Do you understand the connections and dependencies between the parts of your technology, your team, and your market? Are there any conspicuous holes in your network? What isn't covered? What's overcovered?

Within your office walls are there synergies that could be better exploited or ways that members of your team may be inadvertently working at cross-purposes with one another? With the proliferation of Internet service providers, which range from graphics design to expense reporting to document storage, could some tasks be accomplished better and more cost effectively if they were outsourced?

What about the different species in your team? Is your business taking full advantage of their range of capabilities and creativity? This guideline led us to improve our project management process. Some months into the design of one new product, we realized that one of our staff members, who had been focused on research, was quite skilled in a quirky software program that a more senior engineer was struggling with. In another case, the members of a team each assumed that someone else was responsible for contacting a particular vendor. They were all surprised when the critically needed parts didn't arrive when expected. These are communication gaps that can occur in any company but can happen even more easily in a start-up, where people are usually moving fast while wearing a lot of different, and often unfamiliar, hats. By taking a little more time at the outset of projects to overcommunicate and document each of our roles and responsibilities, we revealed gaps and assumptions in our planning as well as reducing miscommunication once the project got under way. Just keeping this principle out in the open allowed us to talk more directly about the ways that we were, or weren't, collaborating.

Gather and Use Energy Efficiently

In nature, survival is absolutely linked to energy efficiency—always extracting the optimum benefit from the minimum effort. While it would be ideal to install solar panels or wind turbines on the roof of every business, many companies don't have that option in their budgets—at least yet. Government incentive rebates are helping to get renewable energy into businesses and homes, but in the meantime any company can take steps to use energy more efficiently. Many utility companies now offer free energy audits, where a technician will review your energy use compared to similar-sized buildings, as well as visit your facility to give you advice on how to improve your efficiency. It's surprising how much energy is lost by most buildings—from poorly insulated roofs, leaky air-conditioning, heater ducting, and lightbulbs that haven't been changed to compact fluorescents (CFLs), to name just a few issues. Most of us are already aware of CFLs and their savings—more than $40 per lightbulb over its lifetime. CFLs have been around since 1976 and readily available since 1995, yet while the European Union, Cuba, and Australia mandate that all new lights be CFLs, only 10 percent of U.S. lightbulbs have been replaced. As the U.S. government's Energy Star program reports, "If every home in the U.S. replaced just one lightbulb with a lightbulb that's earned an Energy Star, the energy savings would light three million homes for a year, save about $600 million in annual energy costs, and prevent 9 billion pounds of greenhouse gas emissions per year, equivalent to those from about 800,000 cars." With statistics like that, the U.S. Department of Energy has gotten onboard with tougher lighting efficiency standards.

Beyond the literal sense of energy efficiency, this guideline is also a reminder that businesses are made of people, translating their life energy into achieving shared goals. Human dynamics in businesses are huge users of energy, so think critically about where you and your staff spend the most time. This could translate into practical decisions like holding more frequent, brief staff meetings, which have been found to be more effective than less frequent, long ones.

Many of us have heard about the "80-20" rule. Popular as a business principle a couple of decades ago, it still offers food for thought today. The "Pareto principle," originally named after an Italian economist who noted that 80 percent of land in Italy was owned by 20 percent of the population, refers to an 80-20 ratio that pops up in many aspects of business and life. For example, most businesses find that 80 percent of their sales come from 20 percent of their contacts—and 20 percent of their sales staff—while 80 percent of complaints come from 20 percent of their clients. This means that we can spend an inordinate amount of resources on a minority of clients (or problems) for little return.

By thinking creatively, any business can get more bang for its energy buck.

Optimize Rather Than Maximize

Looking at the difference between the adolescent period of a ecosystem's development—the maximizing, acquiring, developing stage—and the more mature stage, which focuses on optimizing what you've got, Janine Benyus observes that industry is currently in a state of arrested development: lots of focus on building and acquiring businesses, much less on improving and maintaining them. The lesson is to take a fresh look and focus on quality, not just quantity.

What stage of a business are you in? Whether you're focused on getting investor money and your foot in the door with clients, are in a time-crunched product development team, or work in a long-standing business that wants to keep its market position, here are a few things to think about:

Optimizing time—The concept of seeing time as a valuable resource, much like money or energy, is echoed in a number of these principles. Nature doesn't waste time. Everything is either growing or dying to recycle back into the system. How can you use the time you spend on tasks more effectively, in essence spending less time but achieving more value for it?

Optimizing focus—Nature displays total focus in the predator-prey relationship. How can you grow your company into new areas without becoming distracted? Have you clearly defined who will watch the past and who is looking to the future—and are they communicating with one another? The members of our sales teams self-select who will be responsible for maintaining existing relationships versus hunting new business. We find that people naturally gravitate toward one or the other anyway and tend to unconsciously give less attention to the other type of customer. This way our staff can specialize in meeting the needs of their designated group while doing what they like best.

Optimizing creativity—Nature never stops adapting and evolving. In our rapidly changing world, does your company choose the most creative (productive, innovative, efficient) solution or just the first one that works? Are you strategic or reactive?

Like many companies, we do postmortems after a project wraps up, to see what worked, what didn't, and how we can learn from it. Applying this principle to our discussions revealed that we're much better at seeing when someone else isn't optimizing their time, focus, or creativity than recognizing it in ourselves! If this is true at your company, be a role model for your team. It takes courage to ask your colleagues for suggestions on how to improve, but the reward is worth the risk. Many heads are better than one—until the ultimate decision. Then, remember that you can't command a ship by committee.

Use Materials Sparingly

Nature uses the minimum of materials to get the job done. There are no offcuts. It makes exactly what it needs, when it needs it. This challenge has been adopted by just-in-time manufacturing, and the practice is spreading to other industries. The online shopping giant Amazon is shipping in smaller, lighter packages. Plastic soft drink bottles are getting thinner. Cars have lost weight—about one thousand pounds in the past forty years. And notwithstanding the design and manufacturing complexity, your smartphone is a great example of packing many tools in a very small box.

With just a bit of thought, your company can certainly reduce its use of materials—and save money in doing so. As a simple example, many of us forget that a sheet of paper has two sides. The United States alone uses four million tons of printer paper each year, or seven hundred billion sheets. Using the general rule that it takes seventeen trees to make one ton of copy paper, that's sixty-eight million trees—and while recycling has become far more common in offices and factories, almost 40 percent of paper still ends up in landfill. A sheet of paper saved is also one less that needs to be paid for. A review might find some surprising ways that your firm is using—and paying for—more than it needs.

Let's expand the term *materials* to include any resources that are valuable to us: time, people, creativity, insight. For example, in one of our companies we experimented with reducing out-of-state travel to make introductory presentations to engineering firms and increased our use of webinars—seminars held online that can be attended live or be recorded and listened to at a convenient moment. Time on the road and travel costs went down, as well as our carbon footprint, while the number of potential customers we connected to went up.

Viewed this way, here are some thoughts to consider when attempting to use resources sparingly. Rather than thinking "How many people need to be involved in this task?" ask yourself how *few* could be involved without compromising the task. What's gained with more involvement, and what's lost? Where do we overbuild or overprocess? What motivates the excess? Is it necessary, or is it a leftover practice that's no longer needed? Can we reuse processes—essentially, using one process for a variety of effects? Can you curb your instinct for more and better understand and use what you really need? When your staff insist that they need to hire more people, determine if the new hire will actually improve the bottom line. What are your leverage points? As Archimedes said, you can move the world with the right lever.

Don't Foul Your Nest

This expression paints a clear picture. All organisms breathe, eat, and sleep in their habitat. They can't afford to poison themselves or the ecosystem that supports them, but humans seem to have forgotten this critical survival strategy. The clearest way we can mimic this is to develop products that are nontoxic from the start. Do you know which industrial chemicals are being used in your company? Can you reduce or replace toxic agents in your processes? Are there any other ways that your business has a long-term negative impact on your local environment? If so, how can you remediate them? Fouling your nest can also apply to morale in the workplace. One nationwide firm made this an explicit company value: Employees from clerks to the CEO commit not to say anything about any colleague that they wouldn't say to that person directly.

Don't Draw Down Resources

A good way to remember this principle is that a mature ecosystem lives off the *interest* of its invested life energy, not the principal. A business has to do the same or it goes bankrupt.

Nature doesn't live beyond its means. Wild grazing animals move through a large territory, returning to an area after its plants have regrown. A predator doesn't typically wipe out the entire stock of its prey. In order to be sustainable, a system must use resources at the same rate that they can be regenerated or substituted with benign alternatives. That was easier generations ago, when our economies were based on seemingly infinite natural resources: endless forests, untapped oceans, and mountains of minerals to be discovered. When the global population grew above two billion people, however, we outstripped nature's ability to recoup. In fact, scientists have calculated that we already need the raw materials from one and a half planets the size of earth just to sustain the living conditions we have now. If the emerging world rises to meet the developed world's standards of consumerism, we'll need even more multiples of earth's finite resources. Those of us living in the United States, for example, are already using five planets' worth.

Are there ways that your company is living beyond its means—economically or environmentally? Is your business activity diminishing the viability of the planet? If yes, what can you do as a matter of urgency to correct it? With seven billion of us, any compromise that is unhealthy, no matter how small, adds up to big impact. In the same way, any investment in sustainability diminishes inertia and helps turn the ship of our planet in the right direction.

Remain in Balance with the Biosphere

Our biosphere is the thin layer of air, land, and water that supports life. It is a closed system—the molecules recycle but no new ones are added. At a macro level, this principle relates to objectively informing ourselves about pollution and particularly greenhouse gas emissions. At an industrial level, a balanced biosphere would require that all manufactured items are in a closed loop, with complete recycling and nothing new introduced. That is not the case with our industrial systems, which rely on continuously bringing in new raw materials and results in mountains of offcuts and unused waste. This is a one-way street to full-scale depletion. With nature as a mentor, we can certainly get this right, but if—and only if—we decide to.

Run on Information

In business and in life, having the right information at the right time can mean the difference between life and death. As Janine Benyus explains, in a sustainable ecosystem, "A rich feedback system allows changes in one component of the community to reverberate through the whole, allowing for adaptation when the environment changes." That includes a cascade of chemicals telling your body to heal a wound or one reindeer or wildebeest in a herd signaling danger to hundreds of others. As Janine describes, animals have also evolved exquisite precision in their communications. For a wolf, subtle movements signal the difference between "let's mate" and "you win—I'll leave gracefully." It doesn't work out well if he gets these mixed up.

We learned quickly at PAX how critical it is to develop high-quality, responsive communications. This is part of crossing the divide for a bio-inspired products company, but every business lives and dies on the quality of its information. How quickly—and efficiently—do changes in one area reverberate through the rest of your department or company? How often do you wait for information to come to you, and how often do you seek it out? Do you know who needs your updates—and when you need information, do you know where to go to get it?

Companies can also develop a communication bias that is based on a leader's style, since people tend to have quite strong but often unconscious preferences in how they receive and process information. This mismatch can cost you dearly. I described earlier how helpful it has been at PAX to work with organizational consultants, to understand and improve the ways we communicate. How diverse are your communications channels? How flexible? At PAX we've battled the not invented here syndrome in many industries. How receptive are you and your company to receiving new ideas and information?

Shop Locally

At PAX, we use as many local merchants and service providers as possible. This usually results in better rapport and faster turnaround times. Those vendors that are distant, including software consultants or lawyers, often provide services that don't include significant transportation demands. That feels pretty good, though a second look at the products that we all use every day tells a different story. The next time you sit down to eat, take a moment to consider each of the items on your table. The utensils may have come from China, the glassware from Poland, the wine from France, the water from Fiji, the tomatoes from Mexico, and the chicken from Tennessee. Daily life for most Westerners is populated by the United Nations. That's globalization in action, which keeps economies moving and reduces the incidence of wars. But this long-distance habit is only sustainable if transportation is cheap and reliable, with no accumulation of pollutants or waste. Are there ways you can minimize the amount of transportation and shipping involved in your day-to-day operations? What if something happens to one of your key suppliers? Can you re-create those materials closer to home, or are you dependent on a sole-source supplier? Some wild creatures use a sole source food. Pandas and koalas come to mind; because of the loss of their food supply due to human impact on their habitats, their long-term survival is highly tenuous.

As we've seen, some of these ten principles apply more to some companies than others, but any organization can optimize its impact by learning from nature. Let's meet four that are leading the way.

BEYOND BENIGN

Successful biomimetic companies can be found working with and learning from the biggest redwood trees down to the smallest molecules of industrial chemicals. Green chemistry is a subset of biomimicry that is rapidly gaining supporters worldwide. This discipline copies the way nature performs its infinite variety of chemistry: at ambient temperature, using sunlight and water, and without creating toxic by-products. John Warner is a pioneering "green chemist," who is transforming the world of industrial chemistry. His Warner Babcock Institute for Green Chemistry (WBI) exemplifies how someone can go from concept to creation in building a biomimetic business. Just four years after being launched, the company is profitable and has numerous nontoxic products entering the market. Some might think nontoxic is synonymous with "weak in performance." Not so. Though decades ago some early green cleaning solutions might not have worked as well in the kitchen as the more corrosive, polluting kinds, this is not the issue with modern, well-engineered green chemicals.

WBI is developing super-low-cost solar panels made of chemicals so benign that you can drink them. The company has also developed biodegradable bags that can be made watertight or water soluble, at will, by shining UV light on them; in other words, they can decompose in sunlight if abandoned on a beach or in a field.

John's work springs not only from an interest in nature but also from a deeply personal experience. In the late 1990s, he was a brilliant young industrial chemist who already had two thousand new molecules to his name. Following the death of his two-year-old son to a birth defect, John was haunted by the thought that one of the molecules he'd invented might have caused his son's illness.

As John explains, "There are at least eighty thousand man-made chemicals in use today. Yet there is almost nothing known about the long-term effects or the interactions of these chemicals with one another or their consequences for human health and the environment. Lab chemists have used masks and protective gear as standard practice for decades to develop and handle industrial substances, but in the United

States, fewer than five hundred chemicals have ever been evaluated for toxicity and no more than five have been regulated. There is not a single university in the world that requires a student to take a course in toxicology to become a degreed chemist. Imagine that! In addition, there is virtually no governmental oversight, and the onus is not on manufacturers to prove that their products are safe. Their argument against it? A company could test for years—but how can it test enough to "prove" that a chemical is safe in every possible combination or interaction? Instead, it usually falls on sufferers in the public to prove that their cancer clusters or other ailments are a result of industrial toxins."

The novel, though it should be obvious, premise of green chemistry is that it's better to design safe, clean molecules to begin with. Green chemistry sets the initial requirements for a chemical or chemical process to be functionally equivalent to or outperform existing alternatives, to be more environmentally benign than existing alternatives, and more economically viable than existing alternatives.

You might think that the $3-trillion chemical industry would offer stiff resistance to a radical thinker proposing nothing less than the complete overhaul of its practices. Not so. Unlike some biomimics who come from outside their target markets, John's credentials in industrial chemistry and academia are unbeatable. He speaks his clients' language and understands their priorities.

"I never tell a client, 'We need to be going green,'" John said when we spoke recently. "Companies don't want to hear that. They aren't looking for just aspiration; they need a solution." John and his team create molecules that do the job with a design approach that is more profitable and eliminates the potential liabilities associated with conventional chemicals. His process for getting there just happens to be biomimetic. Nature uses only green chemistry.

There is no question in John's mind that the world's entire inventory of industrial chemicals can be made safe. He describes twelve principles of how to do so in the landmark book *Green Chemistry: Theory and Practice*, which he wrote with one of the EPA's top researchers, Paul T. Anastas.

"Nature is always showing us the best model," explains John. "Because of my training, I know that molecules can be stretched to do something that doesn't suit their fundamental structure, but they'll always strain to go back to where they were before. A biomimetic way of looking at them is more behavioral. Instead of forcing molecules to interact, I 'ask' molecules what their role should be by studying their fundamental structure. For example, some molecules have strong adhesive properties. If it wants to do that, let it be a paint molecule. The benefit is that in chemistry, molecular structure always impacts the manufacturing process, so if you go along with that—like a molecule that already 'knows how' to be a paint molecule—it's going to be a more facile manufacturing route and straightforward product development. We have to let go of ego and let the inherent properties of materials teach us what to do." I understand John's orientation. At PAX, we let fluids in motion show us how they prefer to flow, rather than starting from what an engineer's diagram wants them to do.

John and his partner Jim Babcock started the Warner Babcock Institute four years ago in John's living room. They quickly grew, first to a seven-thousand-square-foot space and then to a forty-two-thousand-square-foot facility with forty scientists. The company has filed more than 250 provisional patents, has a few products already on their way to the market and even more behind them, nearly two dozen clients, and nearly two dozen successes. John is proud that the firm is already cash-flow positive and doubling in size each year, but he hastens to add that his firm isn't succeeding because they're smarter than anyone else. As he says, "How can we be smarter than companies with ten thousand scientists? WBI is outperforming every chemical company on the planet, but it's not that the others aren't motivated. We're just different.

"Their conventional approaches have exhausted what they're trying to do. Biomimicry combined with green chemistry is the best avenue to creativity and innovation."

WBI has a two-pronged business model. The first part is contract invention. Big companies contract the firm to build new molecules, paying fees for development and a success fee or royalties when the new chemicals enter the market. The second aspect of their model is due to John's reach into industry. When he walks the halls of the giant chemical manufacturers with CEOs and vice presidents of research, they tell him,

"I wish we had a molecule that did such and such." WBI reinvests its revenues, or finds grants and outside investment, to research those dream molecules independently. John's goal is to return with a game-changing nontoxic molecule, knock on the CEO's door, and say, Here you go.

With such a successful track record, I asked John if he had any challenges in his work. "Oh, I'm stressed," he said with a rueful laugh. "I'm in charge of business development and I feel the responsibility. Forty people and their families depend on me and my CEO. So I travel a lot and work about eighty hours a week—at times I'm pretty fried." Until WBI has enough molecules in the market to bring a steady flow of royalty revenue back to the company, John is constantly aware of cash flow. He went on to describe what I've heard so many times from biomimetic entrepreneurs. In our society, there just aren't good funding options for research outside of academia. John has had numerous venture capital firms offer to write checks for $20 million or more to get a particular product to market, but the institute isn't looking to grow in that direction. It needs just $500,000 to go from concept to patent, and do that over and over. John confirms that in industrial chemistry, like in the markets that PAX has entered, there's a critical hole in the funding structure. As he says passionately, "While this country can't decide what, if anything, to do, China and India have gotten it together. China has already created fourteen national research labs inventing environmentally benign molecules, all supported by the state. India is mandating an education curriculum for green chemistry. For the first time, a country is requiring that its chemists study toxicology and environmental consequences."

John spent ten years as a university academic, winning such top awards as a scientist and professor that he was invited to meet the president in the Oval Office. As he explained, "The U.S. government and universities do fund academic research, but the currency of academia is not solving problems; it's publishing scientific papers. Studies show that only one in one hundred papers is actually ever read, though—so the standard for success is not producing any benefits to the country. Meanwhile, our student scientists are not being trained at all in the fields of toxicology or green chemistry."

If John had three wishes, he would first mandate that all scientists have training in toxicology—people in the business of making molecules should be able to anticipate harm. He would also require the disclosure of all chemicals in every industrial product. He agrees that the way a chemical is made should be patentable, but when you go to the grocery store, you get to read a label on your food that lists the ingredients. Instead of wasting time arguing about what's toxic, people should have the right to decide for themselves whether or not to buy a product based on the facts. Transparency is the way to change the paradigm. To support that, John's third wish would be the establishment of regional toxicology test centers that would train technicians to conduct low-cost standardized testing of ingredients. States could sponsor these centers, which would help local communities. And by offering lower-cost testing, for example, these centers could also provide financial incentives for companies to disclose their results.

Along with his work at WBI, John is dedicated to his nonprofit for green chemistry education, Beyond Benign. His wife, Dr. Amy Cannon, runs the organization, which is housed in WBI's facilities and hosts a type of art gallery of green solutions as well as town hall meetings on green chemistry and green business. One of Beyond Benign's goals is that 500 out of the 650 U.S. colleges and universities that give chemistry degrees would require at least one course on mechanistic toxicology and environmental mechanisms. To John and his family, work-life balance means doing work that is so satisfying that it gives life meaning—the dividing line between work and real life is nonexistent.

When I asked John Warner if he had advice for aspiring biomimicry entrepreneurs, his answer was prompt. "Bifurcate. Have a passion for whatever you want to call it—environmental sustainability or biomimicry, as well as a passion to do something in the business world. But don't use being green to excuse a lack of science. Sustainability in the absence of traditional standards is a waste of time—it's our biggest enemy. Go out and find the best mentors you can to help on the business side. But mentors in sustainability aren't there yet—so don't lose time looking for them. Just refuse to be unsustainable yourself."

TREEPEOPLE

The city of Los Angeles imports 60 percent of its water each year by pumping it at great cost over mountains from hundreds of miles away. In the same time period, the city deliberately drains to the ocean almost the same amount—115 billion gallons—of local rainwater. Andy Lipkis saw this and asked, What would nature do? The answer is simple: store water in trees and other aquifers.

When it comes to learning from forests, Andy is an old hand. As a teenager in 1973, he started a Los Angeles–based nonprofit to support tree planting at his summer camp. In the years since, TreePeople "has planted more than two million trees in Los Angeles." The nonprofit has won international awards for its groundbreaking work and is the model for using urban forestry to improve urban sustainability. One of their initiatives is the promotion of what TreePeople describes as "Functioning Community Forests." Neighbors join together to make a plan for "an urban forest in their neighborhood"—not separated from their homes in a nearby park but in their own yards and sidewalks.

TreePeople trains citizen foresters to help create living ecosystems in each neighborhood. The fixes are relatively simple and the positive impact on property values, water use, community spirit, and lowered crime rates are significant. Families "plant native, drought-tolerant plants in their yards to reduce the need for irrigation. Downspouts are redirected toward gardens instead of to street gutters. Cisterns and barrels act as aquifers to store water for later use. Trees are planted to create shade and cool streets, walkways, parks, and buildings." City permits allow the introduction of "swales—trenches planted with native vegetation—to slow rainwater and help it soak into the ground," along with "permeable paving that replaces hard asphalt surfaces and lets more rainwater soak into the ground."

Besides providing training and support for tree planting and urban forests, TreePeople operates a conference center and park, conducts extensive outreach education, and runs pilot programs to demonstrate sustainable solutions to urban ecosystem challenges. Andy and his team also advise governments far and near. The organization is a great example of a nongovernmental agency that is impacting municipal policy. A report on sustainability planning for Los Angeles showed that TreePeople's approach of harvesting and better using rain could cut Southern California's water imports by as much as 50 percent—simply by following one of the lessons we can learn from forests—treat waste (rainwater) as a resource.

IN GOOD COMPANY

Interface is the world's largest maker of modular carpet. Its highly respected, pioneering CEO Ray Anderson passed away in 2011. His legacy is Interface, a manufacturing company that grew dramatically by changing its practices to the tenets of biomimicry. In 1994, Ray was looking for inspiration for a speech he needed to give on Interface's environmental position. As Ray often shared, he had what he called a "spear in the chest" epiphany when he read Paul Hawken's book *The Ecology of Commerce.* He dug into the principles of sustainability and biomimicry, began changing Interface's business methods, and asked Janine Benyus, Paul Hawken, and others to join a "dream team" of advisers. Since then, Interface has "increased sales by 66 percent, doubled earnings, and raised profit margins, while cutting its greenhouse gas emissions by 82 percent, fossil fuel consumption by 60 percent, waste by 66 percent, and water use by 75 percent. The company has also invented and patented new machines, materials, and manufacturing processes."

Interface carpet tile

This highly successful company uses nature as a business mentor while it creates recyclable, replaceable, environmentally friendly floor tiles. Their fastest growing line, the Entropy carpet, even looks like the seemingly chaotic patterns found on a forest floor. As designer David Oakey explains, "Each Entropy carpet tile is distinct and varied, but when laid together, they blend into a cohesive pattern." Because the carpet squares have many thin lines of multiple colors, Interface can use a range of dye lots during the production process. And because "the tiles can be set in any direction, there's less waste during installation." The net result is a savings in materials and cost, with attendant reduction in inventory and waste.

Imagine if every company followed this model—greatly increasing wealth while greatly decreasing its negative impacts. Two of Ray's books, *Mid-Course Correction: Toward a Sustainable Enterprise: The Interface Model* and *Confessions of a Radical Industrialist: Profits, People, Purpose—Doing Business by Respecting the Earth,* describe the evolution of Interface. His leadership proved that companies can make more profit by copying nature; in Interface's case this was aided particularly by thinking in new ways about waste reduction and reuse of resources. Now, its "Mission Zero" goal is to run a healthy business with zero negative impact on the environment, which Ray described as the most powerful motivating initiative he'd seen in fifty-five years in business.

Interface now has a $750 million market capitalization value. Its four thousand employees share a sense of higher purpose and have a strong, positive company culture—a source of deep satisfaction to Ray. The company's story debunks the myth that financial success and environmental success are mutually exclusive, yet other companies have so far been slow to follow suit. Ray was, according to those close to him, deeply troubled by this. As a tribute to Ray Anderson in the *Guardian* described, "He referred to himself as a 'recovering plunderer', because he believed that any company that takes more from the planet than it gives back is involved in plunder. As he told a TED conference in 2009, 'Theft is a crime. And the theft of our children's future (will) someday be considered a crime.' Ray forecast that a new generation of CEOs would emerge" to change the business paradigm. He hoped that Interface's success would inspire one individual after another to make different choices.

3.8 BILLION YEARS OF INSPIRATION

There's a wise saying that if you give someone a fish, they eat for a day, but if you teach someone to fish, they eat for a lifetime. Janine Benyus and Dayna Baumeister founded Biomimicy 3.8 to teach the world how to fish for sustainable solutions. Several of their colleagues manage educational programming for Biomimicry 3.8. Their Web site (www .biomimicryinstitute.org) is a rapidly evolving hub and clearinghouse for schools, museums, zoos, and other educational facilities. Teachers can network with one another, share biomimicry curricula or develop lesson plans by using the site's templates and database of strategies, and receive online training including continuing education credits.

As Sam Stier, director of youth education and conservation for Bio-mimicry 3.8 explained, the organization is also developing its own biomimicry curricula on "big topics," like green chemistry, that most individual teachers or school districts might not have the budget to take on independently. The goal is not to replace existing course work but to augment it and give teachers an easy way to bring an engaging new perspective into the classroom. For example, in a typical high school chemistry lab there are lots of Bunsen burners and toxic chemicals— which is a hologram of industrial manufacturing's "heat, beat, and treat" orientation. The institute saw a need for a series of lab-based lessons that illustrate the way nature does chemistry. For example, one of the lessons illustrates how the bio-inspired start-up Calera makes cement without digging quarries, at ambient conditions, and how this removes carbon dioxide from the atmosphere instead of adding to it. Traditional cement production involves heating quarried limestone to 1,400° Celsius, which releases around 5 percent of humanity's annual greenhouse gas emis-sions. Through a hands-on lab using inexpensive materials a teacher can purchase at a grocery store, Biomimicry 3.8's lesson demonstrates that people can make environmentally sustainable concrete the way coral does, by emulating the organism's chemistry.

While biomimicry doesn't yet hold a key position in state or federal education frameworks, the team at the Biomimicry 3.8 Institute is see-ing increased interest from policy makers. Engineering and design have almost equal prominence with science in the latest conceptual frame-work put out by the National Academy of Sciences, and sustainability is finally being referenced by the National Science Foundation and simi-lar institutions. Sam theorizes that the writers of these frameworks and research granting bodies may have been seeing sustainability as more of a political movement than a scientific discipline, so it's taken some time for its academic value to take hold.

In addition to supporting kindergarten through high school educa-tion, Biomimicry 3.8 works with a number of college and university affiliates. Each year, new schools are joining in to offer courses, minors, or even PhDs in biomimicry. Research labs on biomechanics are being established in many universities, either by individual professors who have a passion for a particular field of study or by departments who see important opportunities to transfer academic research into commercial applications. The Center for Interdisciplinary Bio-inspiration in Educa-tion & Research (CiBER) at UC Berkeley; the biomimetics group at the University of Bath; the Ontario College of Arts and Design University; Jilin University and Shandong University in China; and the University of Applied Sciences of Bremen, Germany, are just a few of the institutions who have developed courses, degrees, or departments in biomimicry. A biomimicry robotics research center has been created by École Poly-technique Fédérale de Lausanne. Their robots have replicated the way that grasshoppers, crickets, and flying squirrels leap when they take off, instead of taxiing down a runway. The institute's new jump glider robot jumps up and glides to a distant landing, then jumps again, and so on. The Swedish Center for Biomimetic Fiber Engineering (Biomime) is a multidisciplinary group that offers PhDs to students worldwide and has a special focus on the biosynthesis, self-assembly, structure, and proper-ties of wood fibers in order to optimize this information for advanced materials design.

Sam Stier highlighted one university that is collaborating with Bio-mimicry 3.8 to research human dynamics. The Georgia Institute of Tech-nology has a particular strength in cognitive sciences. Their Center for Biologically Inspired Design (CBID) has been studying how bio-inspired designers and engineers, who often think differently about problems, can best work together. CBID and Biomimicry 3.8 will use the findings to improve Biomimicry 3.8's AskNature.org database of plant and animal success strategies, so that scientists, engineers, and businesspeople can best collaborate to bring research to real life.

Whether you run your business in a biomimetic way, run a business based on biomimetic innovation, or educate the next generation, the benefits are clear. Nature's operating instructions throw open the door to far-reaching opportunities for increased wealth as well as a healthier world.

THE NATURE OF HOPE

Do you remember the unlimited optimism of your childhood, when anything seemed possible? And then we grow up, and sometimes the mess we're in can be pretty overwhelming. It's clear from the body of evidence pouring in that the earth's health is sliding downhill precipitously. It can feel hopeless, and scary, like we're racing the clock; or maybe it's even too late.

Historically, humans have derived power from a number of sources including our own muscles, harnessed animals, forced labor (slaves), water pressure (from watermills to hydropower), solar collection (for drying fruit to heating water to today's solar panels), wind energy (from sails to windmills), and fire (from biomass to fossil fuels to nuclear). These means have served mankind to varying degrees of efficiency. The one billion of us living in the developed world continue to have access to cheap, high-grade energy and the fruits of that energy. But we're learning that the price of this energy is not so cheap, not just in monetary terms but in the unpredictable consequences of massive increases in carbon dioxide emissions and coral-destroying acid buildup in our oceans.

The cost to the environment of our cheap power is overwhelming its benefits and, particularly in the last few decades, systems have been breaking down at a faster and faster rate. At this pace, within the lifetimes of many presently living, societies and our earth will largely cease to function in the ways that make life sweet. From where we are, it looks bleak—and it is—but the game isn't over unless we fail to act.

The greatest wealth and satisfaction of our desires lies not in clear-cutting forests, stripping the oceans, or spewing billions of tons of toxins into the atmosphere but within the forces of nature's wild movement and growth. Nature is the mother of all invention and many of humankind's greatest achievements have been made by copying nature. However, our copies have been rough. We haven't succeeded in mimicking nature's grace, efficiency—and most importantly—sustainability. We're coloring outside the lines and making a mess. Let's look again, using nature as our model as the earliest humans did, but aided by the tools of science.

With nature, it's never too late. Nature is a survivor. Nature never gives up. She heals all wounds. Nature pushes up tiny little blades of grass through city concrete and asphalt and overgrows Mayan cities. She keeps putting out billions of seeds, spores, and baby spiders, growing mountains, evolving new species. She is always creating. It's not just okay to feel optimistic, it's natural, and essential. Combining our human intelligence with optimism is the best way we can give back to our earth. Right now, across the globe, we humans, the products of nature, have the skills and the technology to solve just about any problem we're facing, without sacrifice—if the will is there. There is a way, if we allow ourselves to be guided by nature's optimism and nature's wisdom.

We can do it.

EPILOGUE

I remember sailing tropical seas, about six hundred miles west of northern Sumatra in 1986. This was a lonely part of the Indian Ocean, well removed from shipping lanes. Winds were slight and the water almost smooth as we ghosted along under barely full sails. There had been no sight of land or other boats in almost a week; we five on the sailboat existed alone in a blue vastness.

Something on the horizon ahead caught my eye. Was it a ship? I looked through my binoculars and saw what looked like coconut palms. How could that be? There were no islands charted in these waters. The echo sounder confirmed that we were in very deep ocean and the nearest land was hundreds of miles away. We drew slowly closer. I checked the radar and saw there was indeed something ahead; it wasn't a ship and it had substantial vegetation. As we approached, I saw an island of about three acres in size, with several young coconut palms, the tallest fronds about ten feet high.

We sailed right up to the island's edge and I discovered to my amazement that it was floating. A huge entanglement of nature's flotsam and jetsam was held together by miles of ropes and nets lost and abandoned by Japanese, Indonesian, Taiwanese, and Korean fishermen. There were rain forest logs of enormous size washed down rivers by tropical monsoons, fishing floats and coconuts by the thousand, empty turtle shells, shoring timbers from cargo ships, a largely submerged sea cargo container, bamboo, bottles, boots, plastic bags, rubbish bins, drums, and fragments of ruined dinghies. There were thousands of plastic sandals— all different sizes and color, though, oddly, a disproportionate percentage seemed to be left footed.

Asian fishermen are poorly paid, poorly fed, and spend months at a time on the oceans—calm or wild. The work is raw, hard, and dangerous; there's no room for waste. There was no item in this rubbish suggesting excess in the lives of the litterers. However, my attention was caught by another compelling feature. The island was a floating refuge for myriad crabs, barnacles, small and large fish sheltering underneath, and seabirds. I was struck by this collectivizing of human garbage by nature. Rather than spoiling a pristine ocean environment, which at first was all I could see, nature was using, without discrimination, all materials available to generate new life—an entire, productive, self-sustaining ecosystem.

ABUNDANT NATURE

All wealth historically has been derived from natural resources; and given the bounty of nature, it's been easy to think of it as inexhaustible. Now human creativity and technology has created such powerful technologies that we are pulling out resources and generating waste—from petroleum to ocean fish to pollutants to everything in between—much faster than nature, by herself, can restore them.

To borrow a line from Wall Street, we have to see nature as too big to fail. Like the floating island, if we want to survive and thrive, we must weave the economic, social, and environmental conditions we've created into a platform that will support new growth. The years I've spent in my own businesses and the research I conducted while writing this book have convinced me that biomimicry is a fundamental element in that platform.

Evolution is rooted in crisis, in changing conditions that favor a new generation of those who are best adapted. The best principles of biomimicry would have us make the most of crisis, for it gives birth to new opportunities for humanity. Just as the devastation of wildfire rebirths a forest and war or a holocaust can birth a nation, a crisis wakes us up. It drives us to adapt—to cooperate—to put aside discrimination. It forges new partnerships; transcends petty, divisive acts; and accelerates and spurs creativity. In just four years in World War II, for example, the United States changed from a sleepy, depressed economy into the most powerful nation the world has ever seen.

We're living through some of the greatest economic and environmental crises in history. We're trying to make sense of it and the dross is being burned off. This is no time for short-term fixes; we simply can't just go back to business as usual. We need and are experiencing systemic paradigm changes—leading toward nature's greatest imperative—survival, and the creation of conditions that support further life. We're gathering ourselves, forming new alliances. By embracing biomimicry, innovative scientists, engineers, and businesspeople are creating novel, highly effective, and profitable businesses and products. We're on our way to a new, sustainable industrial revolution that can support and enrich life for all humankind while ensuring the preservation of our exquisite planet.

I could imagine that a hundred years from now historians and anthropologists will describe our moment in history, with the advent of the Internet and social media, as the greatest turning point in the ten thousand-year-long career of "modern" humans. For the first time, ever, virtually all people on earth are becoming instantly connected. Children from every corner of the world have access to comprehensive information and data on almost any subject that can be imagined. This technological revolution is the great democratizer. Despots, governments, corporations, and those who would manipulate for personal advantage and against the common good can be quickly exposed. Popular movements can spring up and grow in minutes. This new network of human intelligence—built on observation, experience, and opinion—is firing like the many interconnected neurons of a single brain, an inextinguishable cascade of possibility whose main purpose, from nature's point of view, is the survival and prosperity of the species and of life itself.

It's often reported that 80 percent of all new businesses fail in the first five years. In some cases, this is a form of natural selection, with survival of the fittest weeding out less adapted technologies or teams. In other cases, the business may be ahead of, or behind, its time and can't find its place in the market. Economic droughts like the recent global financial crisis decrease a company's chance of survival even further. Unfortunately, this is probably going to be true for biomimicry companies as well. Since I started writing this book, several promising bio-inspired firms have closed their doors—the majority due to lack of appropriate funding. In most cases, the technology is still intact and viable, but will need to find a new home. Clearly, there needs to be an evolution of new funding models. At the same time, many more biomimetic companies have been started, and existing firms that have invested in biomimicry research are seeing increasing returns.

Our deeply troubled world can be reinvented through biomimicry. Nature's trillions of solutions throw open the door to far-reaching opportunities for building a better world; rescuing our ailing environment and atmosphere; and giving rise to a powerful, new, sustainable economy. To quote rock musician Tom Petty, "The future ain't what it used to be." No matter who you are, you can be a pioneer and leader in creating a new golden age on earth. A sweet twenty-first century and a third millennium are possible.

Imagine.

It's your life, your world, your opportunity, and your responsibility.

The possibilities are endless.

A c k n o w l e d g m e n t s

....

This book is the result of dedicated work by many people behind the scenes. I offer deep thanks to Steven Scholl, Stephen Sendar, and their outstanding team at White Cloud Press, whose integrity and enthusiasm are matched by their collaborative and innovative ways of working with authors. It was a genuine pleasure to work with them. Whole-hearted thanks go to the fabulous Kasey Arnold-Ince, as well as to Brooke Warner, and Kristin Loberg. Your encouragement and skill helped this first-time writer enter the world of publishing.

To the first readers and contributors of photography, particularly Rachael Bertone and Paula and Harvey Cohen—thank you. And while their day jobs kept PAX running smoothly, Leslie Miller and Karen Dionne provided invaluable help in dealing with my many, many rounds of edits.

The story of PAX would not have happened without the unswerving support of each and every member of our staff and advisors, including Paul Hawken, Amory Lovins, Janine Benyus, Robert Heller, Gordon Rock, Rich Gross, Ron Knott, and Vern Loucks. In particular, I thank Laura Bertone and my "young" engineers: Kim Penney, Onno Koelman, Paul Lees, and Robin Giguere, who have become not only family to me, but biomimicry leaders in their own right. To Tom Gielda, Bruce Webster, Kristian Debus, Peter Fiske, Karen Losee, Jason Oppenheimer, Phil

O'Connor, Kim Morris, and Kasra Farsad; your creativity and professionalism inspire me.

I am deeply grateful to my wife, Francesca Bertone, a master of the business of biomimicry, whose dedication and support made my vision for PAX and this book become reality.

Finally, I acknowledge the heroes working in the wild and in labs around the world who were so generous in sharing their insights with me, and the ultimate hero in all that we do: nature.

INTRODUCTION

3   And, despite living in a period: National Science Foundation. U.S. Doc-
    torates in the 20th Century » Appendix A Detailed Doctoral Fields and
    Demographic Characteristics of Ph.Ds, http://www.nsf.gov/statistics/
    nsf06319/appa.cfm; Mario Cervantes "Scientists and engineers. Crisis, what
    crisis?" OECD Observer No. 240/241, December 2003, http://www
    .oecdobserver.org/news/archivestory.php/aid/1160/Scientists_and_
    engineers.html; UNESCO Institute for Statistics 2001 Report: The State of
    Science and Technology in the World page 17. http://unesdoc.unesco.org/
    images/0013/001318/131841e.pdf.

3   Scientists have already identified more than two million species: Carl Zim-
    mer, "How Many Species? A Study Says 8.7 Million, but It's Tricky" August 23,
    2011 New York Times International Herald Tribune, http://www.nytimes
    .com/2011/08/30/science/30species.html; Carl Zimmer, How Many Spe-
    cies Are There. Discover Magazine, http://blogs.discovermagazine.com/
    loom/2011/08/23/how-many-species-are-there-my-latest-for-the-new-
    york-times/; Environmental Literacy Council: How Many Species are There?
    http://www.enviroliteracy.org/article.php/58.html.

4   From the Greek *bios:* Janine M. Benyus, Biomimicry : Innovation Inspired
    by Nature, September 1997, (ISBN 0-06-053322-6); Biomimicry Institute
    Website: http://biomimicryinstitute.org/about-us/what-is-biomimicry.html.
    Biomimicry Institute Website: http://biomimicryinstitute.org/about-us/
    what-is-biomimicry.html.

5   The industrial revolution was also about: Jay Harman, Designing the Next
    Golden Age, Bioneers Conference Plenary October 15, 2004.

6   For example, the ultraefficient human cardiovascular system: Franklin Institute Website: http://www.fi.edu/learn/heart/vessels/vessels.html.

6   3.8 billion years of trial and error: Biomimicry 3.8 Webpage: Life's Principles: http://biomimicry.net/about/biomimicry/lifes-principles/.

7   When hiking in the Alps: Velco company Website: http://www.velcro.com/About-Us/History.aspx.

8   Think about it: NASA Website: "Ask an Astrobiologist" April 30, 2002 http://astrobiology.nasa.gov/ask-an-astrobiologist/question/?id=143.

CHAPTER 1: THE NEXT INDUSTRIAL REVOLUTION

16   It wasn't until the invention: Jonathan Betts, Rob Ossian's Pirate's Cove Webpage: John Harrison, http://www.thepirateking.com/bios/harrison_john.htm

18   the superfast peregrine falcon . . . almost 200 miles per hour: Earth Sky Webpage: http://earthsky.org/earth/fastest-bird.

18   the soaring leap of a spinner dolphin, which engineering calculations still can't fully explain: Fish, Frank E. "The myth and reality of Gray's paradox: implication of dolphin drag reduction for technology." *Bioinspiration and Biomimetics.* 1 (2006) R17-R25.

19   The widely endorsed Toba catastrophe theory: Ambrose, Stanley. "Late Pleistocene human population bottlenecks, volcanic winter, and differentiation of modern humans." *Journal of Human Evolution* [1998] 34, 623-651.

20   Bio-inspired products often see: Ibid.

20   More than $200 billion: United Nations Environment Programme News Center: "Global Investments in Green Energy Up Nearly a Third to US$211 billion" Thu, Jul 7, 2011 http://www.unep.org/NEWSCENTRE/default.aspx?DocumentID=2647&ArticleID=8805.

21   In 1994, a British organization: Triple bottom line: It consists of three Ps: profit, people and planet. November 17, 2009; The Economist, http://www.economist.com/node/14301663.

21   The San Diego Zoo sponsored: Fermanian Business & Economic Institute Economic Impact Report: Ibid.

21   Following the strong response: "Biomimicry and Economics: The DaVinci Index" A presentation by the Fermanian Business & Economic Institute; August 24, 2011, http://www.pointloma.edu/experience/academics/centers-institutes/fermanian-business-economic-institute/da-vinci-index-biomimicry.

21   As Lynn Reaser, chief economist of FBEI: "Economists unveil new biomim-

icry economic index" James Palen, The Daily Transcript, August 25, 2011. http://www.sddt.com/news/article.cfm?SourceCode=20110825czi.

22 The Da Vinci Index shows that: "Biomimicry and Economics: The DaVinci Index": Ibid.

24 Even the proportions of Egyptian tombs: Aidrian O'Connor, Nature's World: Phi / The Golden Proportion in Culture. 2010, http://www.natures-word. com/sacred-geometry/phi-the-golden-proportion/phi-the-golden-propor-tion-in-culture; Mehmet-Ali Ataç Bryn Mawr Classical Review 2004.09.21 of Corinna Rossi, Architecture and Mathematics in Ancient Egypt, http:// bmcr.brynmawr.edu/2004/2004-09-21.html.

24 Heraclitus theorized: Internet Encyclopedia of Philosophy: http://www .iep.utm.edu/heraclit/.

26 Coincidentally, one of the latest evolutions that is benefiting from some of the same principles: Pinnacle Armor company Website: http://www .pinnaclearmor.com/body-armor/dragon-skin.php.

27 Invented by Michael Kelly: National Archives Website, "Teaching With Documents: Glidden's Patent Application for Barbed Wire", http://www .archives.gov/education/lessons/barbed-wire/index.html; Fencing the Great Plains: The History of Barbed Wire Brochure by National Park Service, U.S. Department of the Interior, Homestead National Monument of America, Beatrice, NE, http://www.nps.gov/home/planyourvisit/upload/Barbed%20 Wire%20Brochure,%20final.pdf.

28 Nature uses several types of camoflage, including: Dr. Carlo Kopp, Monash University Webpage: Information Warfare in Biology Nature's Exploitation of Information to Win Survival Contests, http://www.csse.monash.edu .au/~carlo/infowar-in-biology.html.

28 In the 1940s: Oregon Products Company History: http://www.oregonproducts .com/pro/company/history.htm.

29 Pregnant elephants have been seen: University of Iowa Hospitals and Clin-ics Webpage: Nature's Pharmacy: Ancient Knowledge, Modern Medicine, http://www.uihealthcare.com/depts/medmuseum/galleryexhibits/nature-spharmacy/remedies/wild.html.

29 As evidenced in Neanderthal: Marjorie Murphy Cowan, Plant Products as Antimicrobial Agents, CLINICAL MICROBIOLOGY REVIEWS, 0893-8512/99/$04.0010 Oct. 1999, 564–582, Vol. 12, No. 4 Copyright © 1999, American Society for Microbiology.

29 More than seven thousand compounds: Interactive European Network for Industrial Crops and their Applications Summary Report for the Euro-pean Union 2000-2005, 114, http://www.ienica.net/reports/ienicafinal summaryreport2000-2005.pdf

30   This is fortunate: IENICA Summary Report, 114: Ibid.

30   Quinine, from the bark: "Quinine, an old anti-malarial drug in a modern world: role in the treatment of malaria." Achan J, Talisuna AO, Erhart A, Yeka A, Tibenderana JK, Baliraine FN, Rosenthal PJ, D'Alessandro U. Malaria Journal 2011, 10:144,   http://www.malariajournal.com/content/10/1/144; Ten Facts on Malaria, World Health Organization, April 2012, http://www .who.int/features/factfiles/malaria/en/index.html.

30   Aspirin's principle ingredients were recognized in willow bark by Hippocrates: Jonathan Miner, Adam Hoffhines, "The discovery of aspirin's antithrombotic effects." Texas Heart Institute Journal, Tex Heart Inst J. 2007;34(2):179-86.

30   The indispensable bio-inspired painkillers: Ian Scott, "Heroin: A Hundred-Year Habit" *History Today* Volume: 48 Issue: 6 1998, http://www.historytoday .com/ian-scott/heroin-hundred-year-habit.

30   Incidentally, it's been proposed by some historians: John H. Lienhard, Radio Series *The Engines of Our Ingenuity*: Episode 1037, Rye Ergot and Witches, http://www.uh.edu/engines/epi1037.htm.

30   . . . with up to one hundred thousand: Anne L. Barstow *Witchcraze: A New History of the European Witch Hunts.* (HarperCollins, 1995), 179-181 (Appendix B).

31   Alexander Fleming noticed: Encyclopedia Britannica Online, http://www .britannica.com/EBchecked/topic/209952/Sir-Alexander-Fleming/280655/ Discovery-of-penicillin.

31   Named by *Time* magazine: David Ho, "Bacteriologist Alexander Fleming" *Time*, Monday March 29, 1999.

31   Today, more than seven million: Grace Chai, Pharm.D. Department of Health and Human Services, Public Health Service, Food and Drug Administration, Center for Drug Evaluation and Research, Office of Surveillance and Epidemiology, "Sales of Antibacterial Drugs in Kilograms" Table 1, Part 1: Sales of Antibacterial Drugs by Drug Class and Molecule in Number of Kilograms Sold in Year 2009, November 30, 2010.

31   The Centers for Disease Control: Shehab N, Patel PR, Srinivasan A, Budnitz DS, Emergency department visits for antibiotic-associated adverse events, Clinical Infect Diseases 2008 Sep 15;47(6):735-43, http://cid.oxfordjournals .org/content/47/6/735.full.pdf+html.

CHAPTER 2: GOING WITH THE FLOW

39   The first shamanic drawings: Art Encyclopedia 2012, www.Visual-arts-cork. com, Prehistoric Art Timeline, http://www.visual-arts-cork.com/prehis-

toric-art-timeline.htm. Celtic Spirals Designs, http://www.visual-arts-cork
.com/cultural-history-of-ireland/celtic-spirals-designs.htm.

40 The spiral grew in importance: Numerous books and websites detail the
importance of the spiral in cultural history, including the Spiral Zoom Website: http://www.spiralzoom.com/Culture/PrehistoricArt/PrehistoricArt
.html; Geoff Ward, World Mysteries Website: http://www.world-mysteries
.com/gw_Geoff_Ward_2.htm.

41 In 1863, French philosopher: Ayhan Kursat Erbas, Department of Math
Education, University of Georgia, Athens, GA. MATH 7200-Foundations of
Geometry, December 4, 1999, http://jwilson.coe.uga.edu/emt668/emat6680.
f99/erbas/kursatgeometrypro/golden%20spiral/logspiral-history.html;
Ron Knott, PhD. University of Surrey Webpage: Fibonacci Numbers and
the Golden Section. http://www.maths.surrey.ac.uk/hosted-sites/R.Knott/
Fibonacci/fib.html.

42 The fractal, an important tool: MacTutor Webpage, School of Mathematics and Statistics, University of Saint Andrews, Scotland, July 1999, http://
www-history.mcs.st-andrews.ac.uk/Biographies/Mandelbrot.html.

43 Now, with rapid expansion: Jordan Stanway, The turtle and the robot.
Woods Hole Oceanographic Institution *Oceanus Magazine* Vol. 47, No. 1,
2008, http://www.whoi.edu/cms/files/Oceanus_JP_Stanway_46508.pdf.

43 The cochlea of all mammals: National Institutes of Health National Institute on Deafness and Other Communication Disorders Press Release: "A
New Twist to the Cochlea: Why It's Shaped the Way It Is" March 2, 2006,
http://www.nidcd.nih.gov/news/releases/06/pages/03_02_06.aspx.

43 . . . while the shape of our outer ears: Luigi Gori and Fabio Firenzuoli, Ear
Acupuncture in European Traditional Medicine, eCAM 2007;4(S1)13–16
doi:10.1093/ecam/nem106.

44 Like the Renaissance masters: Stephen Marquardt Beauty Analysis Webpage,
http://www.beautyanalysis.com/index2_mba.htm.

44 Buckminster Fuller wrote: Joe S. Moore, Buckminster Fuller Virtual Institute, http://www.buckminster.info/Ideas/03-TetNatureDNA.htm.

45 Since the start of the industrial revolution: Jeremy Symons, What is causing the climate to unravel? National Wildlife Federation, July 3 2012,
http://blog.nwf.org/2012/07/what-is-causing-the-climate-to-unravel/;
Joe Romm, The biggest source of mistakes: C vs. CO2. Think Progress
.org; March 25, 2008, http://thinkprogress.org/climate/2008/03/25/202471/
the-biggest-source-of-mistakes-c-vs-co2/.

46 All life, and even crystals: Theodor Schwenk, *Sensitive Chaos: The Creation
of Flowing Forms in Water and Air,* (Rudolf Steiner Press, 1965), 47.

47 Our earth travels: NASA Ask an Astrophysicist Webpage: http://imagine
    .gsfc.nasa.gov/docs/ask_astro/answers/971028e.html; Bob King, Buckle up
    for the cosmic roller coaster ride of your life, Posted March 27, 2012, http://
    astrobob.areavoices.com/2012/03/27/buckle-up-for-the-cosmic-roller-
    coaster-ride-of-your-life/.

47 Incidentally, if a person is lost: Jan L. Souman, Ilja Frissen, Manish N.
    Sreenivasa, and Marc O. Ernst, Walking Straight into Circles, Current
    Biology 19, 1538–1542, September 29, 2009 DOI 10.1016/j.cub.2009.07.053,
    http://www.cell.com/current-biology/abstract/S0960-9822(09)01479-1.

48 In 1609, Kepler proved: NASA Ames Research Center Webpage Johannes
    Kepler: His Life, His Laws and Times, http://kepler.nasa.gov/Mission/
    JohannesKepler/.

52 The great names of science: Frank Wilczek, Beautiful Losers: Kelvin's Vor-
    tex Atoms. Nova Webpage: The Nature of Reality, December 29, 2011,http://
    www.pbs.org/wgbh/nova/physics/blog/2011/12/beautiful-losers-
    kelvins-vortex-atoms/.

52 The United States burns: Amory Lovins, A Farewell to Fossil Fuels; Answer-
    ing the Energy Challenge, March/April 2012 Foreign Affairs Webpage,
    http://www.foreignaffairs.com/articles/137246/amory-b-lovins/a-farewell-
    to-fossil-fuels.

59 A July 2011 technical article: John Cermak and John Murphy, Select Fans
    Using Fan Total Pressure To Save Energy, ASHRAE Journal, July 2011,
    http://www.amca.org/UserFiles/file/cermak_AMCA_web.pdf.

CHAPTER 3: CATCHING THE WORLDWIDE WAVE

65 During the midst of the Great Recession: Sustainable Business.com News,
    January 10, 2011, http://www.sustainablebusiness.com/index.cfm/go/news.
    display/id/21691. Walter Yang, 2010—bumper year for big biotech. *Nature
    Biotechnology* Volume: 29, Page: 102 Box 2: Global biotech venture capital
    investment DOI: doi:10.1038/nbt.1777, http://www.nature.com/nbt/jour-
    nal/v29/n2/box/nbt.1777_BX2.html.

66 The people of Denmark: "About 5% of garbage ends up in landfills, com-
    pared with 54% in the United States." Henry Chu, Denmark's green creden-
    tials obscure some unpleasant facts, Los Angeles Times, December 6, 2009,
    http://articles.latimes.com/2009/dec/06/world/la-fg-copenhagen-climate6-
    2009dec06; U.S. Environmental Protection Agency (2009, November),
    Table 29 Municipal Solid Waste Generation, Recycling, and Disposal in the
    United States, Detailed Tables and Figures for 2008, Office of Resource Con-

servation and Recovery, http://www.epa.gov/epawaste/nonhaz/municipal/
pubs/msw2008data.pdf.

66 One rapidly growing international organization: Climate Action in Megacities: C40 Cities Baseline and Opportunities Report June 2011, http://www
.arup.com/Home/Publications/Climate_Action_in_Megacities.aspx.

67 San Diego, California, has established: San Diego Biomimicry BRIDGE:
http://www.sandiegozoo.org/conservation/biomimicry/biomimicry/bio
mimicry_bridge; http://www.sandiegozoo.org/conservation/biomimicry.

67 His Royal Highness The Prince of Wales: http://www.princeofwales.gov.uk/
focus/harmony.

68 The term, coined by Jay Westerveld: Jim Motavalli , A History of Greenwashing: How Dirty Towels Impacted the Green Movement, Daily Finance
Webpage, Posted February 12, 2011, http://www.dailyfinance.com/2011/
02/12/the-history-of-greenwashing-how-dirty-towels-impacted-the-
green/.

68 In 2009, TerraChoice: "Greenwashing is still rampant...Of 2,219 products
making green claims in the United States and Canada, only 25 products
were found to be Sin-free." TerraChoice Report 2009, http://sinsofgreen
washing.org/findings/greenwashing-report-2009/.

69 The Australian government has now: Green marketing and the Trade Practices Act, Australian Competition and Consumer Commission, February
2008, 15, http://www.accc.gov.au/content/index.phtml/itemId/815763.

69 The U.S. Federal Trade Commission: Devika Kewalramani and Richard J.
Sobelsohn, "The Greenwashing domino effect" February 3, 2012 Thom-
son Reuters News & Insight Webpage.,http://newsandinsight.thomson-
reuters.com/Legal/Insight/2012/02_-_February/The_Greenwashing_
domino_effect/.

69 As Norwegian consumer ombudsman, Bente Overli: Bente Overli, Use of
environmental claims in automobile industry, Presentation referencing
"The Consumer Ombudsman's main work on environmental and ethical
marketing", March 23, 2010, http://www.oecd.org/internet/consumerpol-
icy/45118979.pdf; http://www.forbrukerombudet.no/id/11040567.0.

70 In early 2012, the International Organization for Standardization: Commerce International Webpage: Biomimicry: Soon to be a European Standard, March 21, 2012, http://www.actu-cci.com/en/Territories/Europe/
Biomimicry-soon-to-be-a-European-standard.

75 even when confronted with Chinese traders: Gavin Menzies, Annex 10–
Evidence of Chinese Fleets visit to Australia – West Coast, August 18, 2011,
http://www.gavinmenzies.net/Evidence/10-annex-10-evidence-of-chinese-
fleets-visit-to-australia-%E2%80%93-west-coast/.

77   According to Forbes: "This year we've counted 1,226 billionaires, an all-time high. At a record $4.6 trillion, the group's combined net worth is up 2%." Luisa Kroll, Forbes World's Billionaires 2012, Forbes, March 7, 2012, http://www.forbes.com/sites/luisakroll/2012/03/07/forbes-worlds-billion-aires-2012/; Pioneering Study Shows Richest Two Percent Own Half World Wealth, United Nations University World Institute for Development Economics Research (UNU-WIDER), December 5, 2006, http://www.wider.unu.edu/events/past-events/2006-events/en_GB/05-12-2006/.

80   Recently, researchers at the Gran Sasso National Laboratory: CERN Press Release: OPERA experiment reports anomaly in flight time of neutrinos from CERN to Gran Sasso, June 8, 2012, http://press.web.cern.ch/press/pressreleases/releases2011/pr19.11e.html.

80   Kodak similarly dominated: Avi Dan, Kodak Failed By Asking The Wrong Marketing Question, January 23, 2012, Forbes CMO Network, http://www.forbes.com/sites/avidan/2012/01/23/kodak-failed-by-asking-the-wrong-marketing-question/.

CHAPTER 4: SECRETS FROM THE SEA

86   Only about seventy-five shark bites: Oceana Webpage: Shark Attack Statistics, http://oceana.org/en/our-work/protect-marine-wildlife/sharks/learn-act/shark-attack-statistics.

89   As one example, the new cruise ship: Jotun Press Release: A Breakthrough in the Fouling Release Coatings Technology March 17, 2010. http://www.specialchem4coatings.com/news-trends/displaynews.aspx?id=13180.

89   there are fifty thousand large ships: Shipping and World Trade: Number of ships (by total and trade), Shipping Facts, October 31, 2010, http://www.marisec.org/shippingfacts/worldtrade/number-of-ships.php.

90   Dr. Anthony Brennan: Sharklet company Website: http://www.sharklet.com/technology/.

90   A vessel can consume up to 40 percent: Fouling, Marine Paint Research for Sustainable Solutions in Marine Antifouling, March 2012, http://www.marinepaint.se/program/marinepaint/marinepaint/fouling.4.61632b5e117dec92f47800078750.html.

91   Dr. Brennan's breakthrough came: Sharklet company Website: http://www.sharklet.com/technology/.

91   In development since winning: Sharklet company Website: http://www.sharklet.com/about-us/faqs/.

92   including recently being awarded a grant: Sharklet company Website: http://

www.sharklet.com/2011/03/sharklet-technologies-awarded-1-2-million-phase-ii-nih-research-grant-for-further-development-of-sharklet-patterned-urinary-catheter/.

92   SkinzWraps, Inc.: Skinzwraps company Website: http://www.skinzwraps.com/.

92   The company's newest product: Fastskinz Webpage: http://fastskinz.com/.

93   The popular company's Fastskin: Speedo company Webpage: About Speedo, http://www.speedousa.com/helpdesk/index.jsp?display=corp&subdisplay=about.

93   Made of a patented textile: Janet Bealer Rodie, "UltraTech, UltraSpeed" Textile World Website, May 2008, http://www.textileworld.com/Articles/2008/May_2008/Departments/QFOM.html.

93   The international organization for competitive swimming: FINA Dubai Charter adopted March 14, 2009, http://www.fina.org/H2O/docs/PR/the%20dubai%20charter.

94   In another shark-inspired technology: BioPower Systems company Website: http://www.biopowersystems.com/.

94   Today, hydroelectric plants account for: Use and Capacity of Global Hydropower Increases, Worldwatch Institute, 2011, http://www.worldwatch.org/node/9527.

95   Brooke Flammang of Harvard University: Elizabeth Pennisi, "How Sharks Go Fast" ScienceNOW Web page, November 29, 2011, http://news.sciencemag.org/sciencenow/2011/11/how-sharks-go-fast.html.

95   After extensively analyzing fish propulsion: Triantafyllou, Michael S. and Triantafyllou, George S. An efficient swimming machine. Scientific American; Mar95, Vol. 272 Issue 3, 64, http://web.mit.edu/towtank/www-new/Papers/efficient-swimming.pdf.

95   BioPower applied the principles: BioPower Systems Webpage: http://www.biopowersystems.com/biostream.html; News page: BioPower Systems receives new funding of $5.6 million from the Australian Government, July 4, 2012, http://www.biopowersystems.com/news.html.

97   . . . can communicate with their relatives: Cornell University, Secrets Of Whales' Long-distance Songs Are Being Unveiled, ScienceDaily March 2, 2005, http://www.sciencedaily.com/releases/2005/02/050223140605.htm.

97   Through a surprising and somewhat humbling experience: Stephen Dewar: in discussion with the author, November 14, 2011.

98   As a fluid dynamics researcher: WhalePower company Website: http://www.whalepower.com/drupal/?q=node/1.

102   How these large land mammals evolved: Understanding Evolution Web-

page: The evolution of whales, http://evolution.berkeley.edu/evolibrary/ article/evograms_03; Getting A Leg Up On Whale And Dolphin Evolution: New Comprehensive Analysis Sheds Light On The Origin Of Cetaceans, ScienceDaily, September 25, 2009, http://www.sciencedaily.com/ releases/2009/09/090924185533.htm.

104   Now, thanks to research by a team: Baleen Filters Website: http://www .baleenfilters.com/; AskNature.org: Baleen filters, http://www.asknature .org/product/19cf015cb42875f96279d581b9f66e35.

105   the filter-feeding gray whale has an enormous tongue: AskNature.org: Lingual rete precools blood: gray whale, http://www.asknature.org/strategy/ 881ce65b3adf9c780f487968c1f056b9; Heyning, J.E.; Mead, J.G. 1997, Thermoregulation in the mouths of feeding gray whales, Science, 278(5340): 1138-1139.

105   at up to $50,000 for each procedure: United Nations Environment Programme Press Release: Entrepreneurs of the Natural World Showcase Their Groundbreaking Solutions to the Environmental Challenges of the 21st Century, May 28, 2008, http://www.unep.org/Documents.Multilingual/ Default.Print.asp?DocumentID=535&ArticleID=5816&l=en.

109   As many as twenty-five thousand: Save Japan Dolphins Website: http://save japandolphins.org/take-action/frequently-asked-questions.

109   In fact, a dolphin's whistles, pulses, and clicks: Michael Stocker, in discussion with the author, October 2004. For more information on bioacoustics, see Ocean Conservation Research Website: http://ocr.org/.

109   A scientist at Penn State's Center: AskNature.org Webpage: Chirps carry through water: dolphins, http://www.asknature.org/strategy/98d95c82f86 e1b1fb819e4fe3f8ed146; Penn State Department of Electrical Engineering Center for Information and Communication Technology Research, http:// cictr.ee.psu.edu/.

110   Using a different aspect of dolphin sounds: Evologics Website: http://www. evologics.de/en/products/index.html.

110   Some species of dolphin: R. Aidan Martin, Maximum Travel Speed of Selected Marine Life, http://www.elasmo-research.org/education/topics/ r_haulin'_bass.htm; Leisurevolution Webpage: Dolphin information by the numbers, June 2010, http://www.leisurevolution.org/the-dolphins/more-dolphin-information/dolphin-information-by-the-numbers.

111   The dolphin's powerful tail: Lunocet company Website: http://www.luno-cet.com/.

111   Another biomimic, Bob Evans: Force Fin company Website: http://www .forcefin.com/rd_navy.html.

CHAPTER 5: SCALES AND FEATHERS

114   an example of biotherapy, where a living animal is used: BioTheraputics, Research, and Education Foundation Website: http://www.bterfoundation .org/.

115   Unfortunately, in part due to their disruption: Mayo Clinic Deep Vein Thrombosis Webpage: Warfarin side effects: Watch for interactions, http:// www.mayoclinic.com/health/warfarin-side-effects/HB00101.

115   A specialist in particle-fluid transport: FBEI San Diego Zoo Biomimicry Economic Impact Study, 18, http://www.sandiegozoo.org/images/uploads/Bio mimicryEconomicImpactStudy.pdf; http://www.companiesintheuk.co.uk/ ltd/brinker-technology.

116   Researchers at the University of Utah: Henry Fountain, New York Times Global Science Material World Webpage: Studying Sea Life for a Glue That Mends People, April 12, 2010, http://www.nytimes.com/2010/04/13/ science/13adhesive.html?pagewanted=all.

119   Scientists are working to unlock: Eve Jacobs, Of Diamond and Dragons, University of Medicine and Dentistry of New Jersey Magazine, Spring 2005, http://www.umdnj.edu/umcweb/marketing_and_communications/publi cations/umdnj_magazine/spring2005/features/09diamondragons.htm.

119   Eli Lilly has synthesized: Eli Lilly Byetta Webpage: http://www.byetta.com/.

120   The former has a routine where: AskNature.org: Stance and skin channels harvest rainwater: Texas horned lizard, http://www.asknature.org/strategy/ e56a43d8621c18f0cbce032ccac6dccc.

120   The thorny devil: AskNature.org: Grooves gather water: Thorny devil, http://www.asknature.org/strategy/3f47ec0d4343c94f82e19e103ac20c34.

120   Evolutionary biologist Dr. Ingo Rechenberg: Ingo Rechenberg and Abdullah Regabi El Khyari, The Sandfish of the Sahara: A Model for Friction and Wear Reduction, Bionik und Evolutionstechnik-Technische Universität Berlin, http://www.bionik.tu-berlin.de/institut/safiengl.htm.

121   My gecko sighting reminded me: The University of California Berkeley's Biologically Inspired Synthetic Gecko Adhesives Webpage includes a link to other research on gecko adhesives: http://robotics.eecs.berkeley.edu/~ronf/Gecko/ index.html; http://robotics.eecs.berkeley.edu/~ronf/Gecko/links.html.

122   Researcher Alexis Debray: Alexis Debray, Manipulators inspired by the tongue of the chameleon, Biomimetics and Bioinspiration, Bioinspir. Biomim. 6 026002 doi:10.1088/1748-3182/6/2/026002; William Weir, The Artificial Chameleon Tongue is Here, April 20, 2011., Hartford Courant Website: http://articles.courant.com/2011-04-20/health/hc-weir-tongue-0421-20110420_1_tongue-acceleration-bitly.

124   Sea snakes breathe air: Scott D. Fell, eMedicine Health Webpage: Sea Snake Bite, http://www.emedicinehealth.com/wilderness_sea_snake_bite/article_em.htm.

125   It's not that the tree snake flies: Jake Socha, Flying Snake Webpage: http://www.flyingsnake.org/. Marc Kaufman, DOD tries to uncover secret of flying snakes, The Washington Post November 22, 2010, http://www.washingtonpost.com/wp-dyn/content/article/2010/11/22/AR2010112206308.html.

125   These pits on each side of the head: AskNature.org: Receptors create thermal image: pit viper, http://www.asknature.org/strategy/c90d536656b2e9166dff8946678c350d; USAF Materiel Command: Delivering War-Winning Capabilities on Time and on Cost, Defense AT&L Interview: May-June 2005, http://www.dau.mil/pubscats/PubsCats/atl/2005_05_06/Mar_mj05.pdf.

127   There is a species of hummingbird: Jean Lee Habenicht, Nectar Art Prints Webpage: Hummingbird migration, http://www.nectarartprints.com/hb_migration_rt.htm.

128   A Boeing 747 airplane: Bernard Choi, 747-8's Million Pound Take-off, Boing Webpage, August 23, 2010, http://www.boeing.com/Features/2010/08/bca_one_million_08_23_10.html; 747 Fun Facts, http://www.boeing.com/commercial/747family/pf/pf_facts.html; Seth Lehman, Fitch Affirms San Francisco Airport (SFO Fuel Co.), CA Special Facil Revs at 'BBB+', Outlook Stable, Fitch Ratings, May 3, 2010, http://www.businesswire.com/news/home/20120503007009/en/Fitch-Affirms-San-Francisco-Airport-SFO-Fuel.

128   Eiji Nakatsu: AskNature.org: Beak provides streamlining: common kingfisher, http://www.asknature.org/strategy/4c3d00f23cae38c1d23517b6378859ee; AskNature.org: The wings of owls allow for silent flight thanks to several types of specialized feathers, http://www.asknature.org/strategy/938e8c4d8e2bf786fa5c9922d181273e.

129   A woodpecker, however: Corey Binns, Why Woodpeckers Don't Get Headaches, January 8, 2007 Live Science Website, http://www.livescience.com/9502-woodpeckers-headaches.html.

129   Sang-Hee Yoon and Sungmin Park: Sang-Hee Yoon and Sungmin Park, A mechanical analysis of woodpecker drumming and its application to shock-absorbing systems, January 17, 2011, BIOINSPIRATION & BIOMIMETICS Bioinsp. Biomim. 6 (2011) 016003 (12pp), http://iopscience.iop.org/1748-3190/6/1/016003/pdf/bb11_1_016003.pdf.

129   As you can imagine: Rebecca Boyle, Woodpeckers' heads inspire new shock-absorbing systems for electronics and humans, Popular Science Online: PopSci Technology Group February 4, 2011, http://www.popsci.com/

technology/article/2011-02/woodpecker-heads-inspire-new-cushioning-systems-electronics-and-humans.

130 Large and black: The Travel Almanac Website: http://www.thetravelalmanac.com/lists/birds-speed.htm.

132 Incidentally, contrary to a popular myth: Dr. Karl's Great Moments in Science: Ostrich Head in Sand, ABC Science. Karl S. Kruszelnicki Pty Ltd, November 2, 2006, http://www.abc.net.au/science/articles/2006/11/02/1777947.htm; wikiHow Webpage: How to survive an Encounter with an Ostrich, http://www.wikihow.com/Survive-an-Encounter-with-an-Ostrich.

133 Scientists at Pennsylvania State University: Penn State Live Webpage: Airplane wings that change shape like a bird's have scales like a fish, April 20, 2004, http://live.psu.edu/story/6497.

133 Boeing is taking a firm step: Morphing winglets under development promise fuel savings and reduced aircraft noise on landing, Green Air Online, February 16, 2009, http://greenaironline.com/news.php?viewStory=374.

133 As scientists at NASA Langley Research Center: Emilie J. Siochi, John B. Anders, Jr., David E. Cox, Dawn C. Jegley, Robert L. Fox, and Stephen J. Katzberg, Biomimetics for NASA Langley Research Center "Year 2000 Report of Findings From a Six-Month Survey," Langley Research Center, Hampton, Virginia, February 2002; Anna-Maria R. McGowan, Anthony E. Washburn, Lucas G. Horta, Robert G. Bryant, David E. Cox, Emilie J. Siochi, Sharon L. Padula, and Nancy M. Holloway, "Recent Results from NASA's Morphing Project," ntrs.nasa.gov/archive/nasa/casi.ntrs . . . /20020063576_2002101691. pdf.

134 NASA, Boeing, and the U.S. Air Force: Harun Yahya, Biomimetics: Technology Imitates Nature, http://harunyahya.com/en/books/3864/Biomimetics_Technology . . ./4504.

134 In the 1970s and 1980s: Biomimetics for NASA Langley, 11: Ibid.

134 This approach was also used: Langley Research Center Fact Sheet: NF203 NASA RIBLETS FOR STARS & STRIPES, October 1993, http://www.nasa.gov/centers/langley/news/factsheets/Riblets.html.

135 Dr. Geoffrey Lilley: John Roach, Owls' Silent Flight May Inspire Quiet Aircraft Tech, National Geographic News, December 17, 2004, http://news.nationalgeographic.com/news/2004/12/1217_041217_owl_feathers.html.

135 As early as the 1960s: Biomimetics for NASA Langley, 12: Ibid.

135 Southwest Airlines: Aviation Partners Boeing Website: Flying further for less: Blended Winglets and their benefits, Airline Fleet & Network Management-March/April 2006, http://www.aviationpartnersboeing.com/pdf/news/AF42_APB.pdf.

CHAPTER 6: THE BEE'S KNEES

138   But maggots have also been used: Whitaker IS, Twine C, Whitaker MJ, Welck M, Brown CS, Shandall A, Larval therapy from antiquity to the present day: mechanisms of action, clinical applications and future potential, Postgraduate Medical Journal [2007, 83(980):409-413], http://ukpmc.ac.uk/articles/PMC2600045?pdf=render.

138   Their published results: Already sourced

138   Maggots also secrete allantoin: Maggot Therapy, 24: Ibid.

138   The blowfly maggot has skin: AskNature.org: Maggot skin strengthens as it dries: blowfly, http://www.asknature.org/strategy/1c314745a494e8807c6941 a4b3284dde.

139   Researchers at the Evolutionary Biomaterials Group: AskNature.org: Capillary action aids adhesion: European blowfly, http://www.asknature.org/strategy/fa1148883f944d31eceda164647bcb2c.

139   The feet of blowflies: AskNature.org: Feet taste food: blowfly, http://www.asknature.org/strategy/192e17022dd44aaffca437409fe2c31f.

139   Researcher Michael Dickinson: Joe Palca, Flies In Danger Escape With Safety Dance NPR Morning Edition August 29, 2008, http://www.npr.org/templates/story/story.php?storyId=94070264.

141   A German company, Honey-Med: Honey-Med company Website: http://www.honey-med.com/shop/katalog/german/Katalog.pdf.

141   Beeswax has also been used: Southeast Texas Honey Company Webpage: Beeswax Uses, http://www.texasdrone.com/Beeswax-Uses.htm.

142   The third century geometer: D'Arcy Wentworth Thompson, *On Growth and Form,* (Cambridge University Press, 1992), 108.

142   A honeybee has to ingest: West Mountain Apiary Webpage: Inside the Hive, http://westmtnapiary.com/inside_the_hive.html.

142   The New York-based company Panelite: Panelite company Website: http://www.e-panelite.com/.

143   A Norwegian firm: Various Architects Website: http://variousarchitects.no/project/mobile_performance_venue/.

143   The Sinosteel skyscraper: Mike Chino, MAD Architects Honeycomb Skyscraper, August 6, 2008, Inhabit.com, http://inhabitat.com/mad-sinosteel-plaza/.

143   Honeycomb walls also inspired: Mike Chino, Housing Complex in Slovenia is a Series of Honeycomb Modular Apartments, February 1, 2011, Inhabit.com, http://inhabitat.com/slovenias-gorgeous-honeycomb-housing-complex/.

143 Nissan Motor Company: Nissan Sustainability Report 2012, 42, http://www.nissan-global.com/EN/DOCUMENT/PDF/SR/2012/SR12_E_P041.pdf; Nissan Press Release: Crash Avoidance Robotic Car Inspired by Flight of the Bumblebee, September 26, 2008, http://www.nissan-global.com/EN/NEWS/2008/_STORY/080926-01-e.html; Nissan Press Release: Nissan EPORO Robot Car "Goes to School" on Collision-free Driving by Mimicking Fish Behavior, October 1, 2009, http://www.nissan-global.com/EN/NEWS/2009/_STORY/091001-01-e.html.

144 Software is a natural place to apply: Tyler Hamilton, Managing Energy with Swarm Logic, MIT Technology Review, February 4, 2009, http://www.technologyreview.com/news/411881/managing-energy-with-swarm-logic/.

145 Wolfgang Sturzl and his colleagues: Colin Barras, Artificial bee eye could improve robotic vision, New Scientist August 6, 2010, http://www.newscientist.com/article/dn19272-artificial-bee-eye-could-improve-robotic-vision.html.

145 Researchers, including Marie Dacke: P. Kraft, C. Evangelista, Marie Dacke, T. Labhart, M. V. Srinivasan, Honeybee navigation: following routes using polarized-light cues, Philosophical Transactions of the Royal Society B-Biological Sciences, March 2011, doi: 10.1098/rstb.2010.0203 Phil. Trans. R. Soc. B 12 March 2011 vol. 366 no. 1565, 703-708.

145 CSIRO has manufactured: Mark Peplow, Insects tapped for better rubber, October 12, 2005, *Nature* Online, doi:10.1038/news501010-9; CSIRO Outcomes Webpage: Resilin project clocks near perfect 98 per cent, 12 September 2008 (Updated 14 October 2011), http://www.csiro.au/Outcomes/Materials-and-Manufacturing/Innovation/ResilinResilience.aspx.

145 In fact, at five times that of steel: Yuka Yoneda, Scientists Unlock Secret of Super Strong Spider Silk Material, May 18, 2010, Inhabit.com, http://inhabitat.com/scientists-unlock-secret-of-spider-silk-which-is-5-times-stronger-than-steel/.

146 A research team at Oxford University: Oxford Silk Group Webpage: http://users.ox.ac.uk/~abrg/spider_site/index.html.

146 It's been calculated that: Glaswerke Arnold Ornilux Webpages: http://www.ornilux.de/cms.asp?Sprache=en; http://www.ornilux.com/history-research.html.

148 Ten million degrees Celsius: George Beccaloni, Blattodea Culture Group Webpage: Cockroaches: An Amazing Diversity, http://blattodea-culture-group.org/content/cockroaches-amazing-diversity;Dr.Karl'sGreatMoments in Science: Cockroaches and radiation, ABC Science, Karl S. Kruszelnicki Pty Ltd, February 23, 2006, http://www.abc.net.au/science/articles/2006/02/23/1567313.htm?site=science/greatmomentsinscience.

149   While conducting research at Nottingham University: Society for General Microbiology, "Insect brains are rich stores of new antibiotics" ScienceDaily, September 7, 2010, http://www.sciencedaily.com/releases/2010/09/100906202901.htm. Bill Hendrick, Brains of Cockroaches and Locusts Contain Substances Lethal to Bacteria, WebMD Health News, September 9, 2010, http://www.webmd.com/skin-problems-and-treatments/news/20100909/cockroach-brains-vs-mrsa; William J. Bell, Louis M. Roth, Christine A. Nalepa, *Cockroaches : ecology, behavior, and natural history,* (The Johns Hopkins University Press, 2007), 78, http://www.cremm.es/ARTICULOS/Cockroaches%20-%20Ecology,%20Behavior%20and%20Natural%20History.pdf.

149   Scientist have been trying for decades: US News.com Science Webpage: Cockroach Inspires Robotic Hand to Get a Grip. January 10, 2011, http://www.usnews.com/science/articles/2011/01/10/cockroach-inspires-robotic-hand-to-get-a-grip.

149   More than twenty-five years ago, Bob started studying: University of California Berkeley Poly-Pedal Lab: http://www.berkeley.edu/news/media/releases/2002/09/rfull/rfull.html.

150   Here's a surprise: Natural History Museum (London) Webpage: Termites are cockroaches after all, April 11, 2007, http://www.nhm.ac.uk/about-us/news/2007/april/news_11364.html.

150   While the outside environment: Abigail Doan, Green Building in Zimbabwe Modeled After Termite Mounds, December 10, 2007, Inhabit.com, http://inhabitat.com/building-modelled-on-termites-eastgate-centre-in-zimbabwe/.

150   States like California: Consumer Energy Center Webpage: Central HVAC, California Energy Commision, http://www.consumerenergycenter.org/home/heating_cooling/heating_cooling.html; Flex Your Power Webpage: HVAC System, Efficiency Partnership, 2012, http://www.fypower.org/com/tools/products_results.html?id=100124.

151   There are now over one billion vehicles: John Sousanis, World Vehicle Population Tops 1 Billion Units, Wards Auto, August 15, 2011, http://wardsauto.com/ar/world_vehicle_population_110815; Joyce Dargay, Dermot Gately and Martin Sommer, Vehicle Ownership and Income Growth, Worldwide: 1960-2030, http://www.econ.nyu.edu/dept/courses/gately/DGS_Vehicle%20Ownership_2007.pdf.

151   Congestion researcher Dirk Helbing: New Scientist Webpage: Ants have a simple solution to traffic congestion, November 6, 2008, Magazine issue 2681, http://www.newscientist.com/article/mg20026815.400-ants-have-a-simple-solution-to-traffic-congestion.html.

154   Butterflies have fascinated humans: Rachel Asher, How to draw butterfly wings,eHowWebpage:http://www.ehow.com/how_5075398_draw-butterfly-wings.html; Obsession with Butterflies Webpage: Frequently Asked Questions and Answers About Butterflies, http://www.obsessionwithbutterflies.com/butterflies.html.

154   There may be as many as twenty-eight thousand species: Papaleng, ScienceRay.com Webpage: Interesting facts about butterflies, May 26, 2009, http://scienceray.com/biology/zoology/interesting-facts-about-butterflies/. Opler, Paul A., Kelly Lotts, and Thomas Naberhaus, coordinators. 2012. Butterflies and Moths of North America. http://www.butterflies andmoths .org/; http://www.butterfliesandmoths.org/faq/how-long-do-butterflies-and-moths-live

154   The brilliant colors on the wings of many butterflies and some bird feathers: AskNature.org: Wing scales diffract and scatter light: Morpho butterflies, http://www.asknature.org/strategy/1d00d97a206855365c038d57832ebafa.

154   Qualcomm has copied: Mirasol Product Website: http://www.mirasol displays.com/.

155   Japanese researchers at Teijin Fibers: Japan: Teijin Fibers Increases "Morpho-tex" Capacity, July 1, 2004, DAISEN Ltd 2008, http://www.highbeam.com/doc/1P3-665959151.html

155   JDS Uniphase: JDS Uniphase Chromaflair Webpage: http://www.jdsu.com/en-us/Custom-Color-Solutions/Products/chromaflair-pigment/Pages/default.aspx; Chromaflair L-Spec Light Interference Pigments Datasheet June 2010, http://www.jdsu.com/ProductLiterature/chromaflairl_ds_dec_ae.pdf.

156   If fully adopted: Numerous case studies and online articles are available that cite at least 80 percent energy savings, with LEDs expected to become even more efficient.

156   But the light emitted: BBC News Website: LEDs work like butterflies' wings, November 18, 2005, http://news.bbc.co.uk/go/pr/fr/-/1/hi/sci/tech/4443854 .stm.

156   Butterflies also benefit: AskNature.org: Wing surface self-cleans: Morpho butterfly, http://www.asknature.org/strategy/f7044d096233ab3467a75d1337 fd52ad.

156   Harvard University is developing: Laura Hibbard, 'Bed of Nails' Surface Prevents Ice Formation, December 16, 2010, Discovery News, http://news .discovery.com/tech/surface-prevents-ice-formation.html; The Aizenberg Biomineralization and Biomimetics Lab Webpage: http://www.seas.harvard .edu/aizenberg_lab.

157   Mosquito is Spanish for: Craig Freudenrich, How Mosquitoes Work, How

Stuff Works Webpage: http://science.howstuffworks.com/environmental/life/zoology/insects-arachnids/mosquito.htm.

157   When receiving an injection: Jerry Emanuelson, Some preliminary information: The Prevalance of Needle Phobia,http://www.needlephobia.com/prevalence.html.

157   No matter how thin: AskNature.org: Mosquito inspired microneedle, http://www.asknature.org/product/b878d9e04cc92c53fbf90c3d855e052b.

158   rapidly expanding number of diabetics: Centers for Disease Control and Prevention Press Release: Number of Americans with Diabetes Projected to Double or Triple by 2050, October 22, 2010, http://www.cdc.gov/media/pressrel/2010/r101022.html.

158   Disease-carrying mosquitos: World Health Organization Fact Sheet 214: Malaria, Fact sheet N°94 April 2012, http://www.who.int/mediacentre/factsheets/fs094/en/; Investing in Global Health Research: Malaria, Global Health Initiative Fact Sheet, April 2007, http://www.familiesusa.org/issues/global-health/malaria.PDF.

159   The malaria-spreading Anopheles mosquito: Stefan Anitei, Killer Spiders That Prey Only Blood Full Malaria Mosquitoes, January 13, 2007, Softpedia Webpage: http://news.softpedia.com/news/Killer-Spiders-That-Prey-Only-Blood-Full-Malaria-Mosquitoes-44364.shtml.

159   Male mosquitos were thought to be: Harmonic Convergence in the Love Songs of the Dengue Vector Mosquito, Science, February 20, 2009, 323(5917): 1077–1079, doi:10.1126/science.1166541, http://www.ncbi.nlm.nih.gov/pmc/articles/PMC2847473/pdf/nihms176530.pdf; Eric Bland, Mosquito Buzz Actually a Love Song, Discovery News, January 8, 2009, http://dsc.discovery.com/news/2009/01/08/mosquito-flap-mating.html.

159   One bug has worked out: AskNature.org: Water vapor harvesting: Namib desert beetle, http://www.asknature.org/strategy/dc2127c6d0008a6c7748e4e4474e7aa1; Radhika Seth, Beetle Juice Inspired! Yanko Design Website: http://www.yankodesign.com/2010/07/05/beetle-juice-inspired/.

160   There is a semi-aquatic rove beetle: AskNature.org: Reducing surface tension to travel on water: rove beetle, http://www.asknature.org/strategy/51b862c4f828ffd2cd5e9c6b9cc47e5f/.

161   Half of all vaccines: Calling the Shots, Manufacturing Chemist, March 1, 2006, http://www.manufacturingchemist.com/technical/article_page/Calling_the_shots/37189.

162   A new species of bacteria: New species of bacteria found in Titanic 'rusticles', BBC News (Science and Environment) January 6, 2010, http://www.bbc.co.uk/news/science-environment-11926932.

162   The evocatively named poo-gloo: Wastewater Compliance Systems company
      Website:http://www.wastewater-compliance-systems.com/index.html;http://
      www.wastewater-compliance-systems.com/municipal.html.

CHAPTER 7: SPORES AND SEEDS

164   Did you know that the world's largest: Craig Schmitt and Michael Tatum, The
      Malheur National Forest: Location of the World's Largest Living Organism,
      http://www.fs.usda.gov/Internet/FSE_DOCUMENTS/fsbdev3_033146.pdf.

165   His talk on the six ways that mushrooms: In conversation with the author,
      December 4, 2011.

172   The world currently produces: Goldsheet Mining Directory—World Gold
      Production, http://www.goldsheetlinks.com/production2.htm; Nina Shen
      Rastogi, Production of gold has many negative environmental effects, Wash-
      ington Post, September 21, 2010, http://www.washingtonpost.com/wp-dyn/
      content/article/2010/09/20/AR2010092004730.html.

173   For example, Chile: Debaraj Mishra and Young-Ha Rhee, Current Research
      Trends of Microbiological Leaching for Metal Recovery from Industrial
      Wastes, Formatex 2010, www.formatex.info/microbiology2/1289-1296.pdf.

174   Each year, just fifteen cargo ships: John Vidal, Health risks of shipping pol-
      lution have been 'underestimated', Guardian, April 9, 2009, http://www
      .guardian.co.uk/environment/2009/apr/09/shipping-pollution; Paul Evans,
      Big polluters: one massive container ship equals 50 million cars, Giz-
      mag, April 23, 2009, http://www.gizmag.com/shipping-pollution/11526/;
      SwizStick, Understanding Bunker Fuel: is there a relationship to crude
      prices? 3PLWire, December 29, 2011, http://www.3plwire.com/2011/12/29/
      understanding-bunker-fuel-is-there-a-relationship-to-crude-prices/.

174   Researchers at the Biomimetics: Darren Quick, Seeds inspire new artifi-
      cial anti-fouling surface, GizMag, July 3, 2011, http://www.gizmag.com/
      seed-inspired-anti-fouling-surface/19094/.

174   The global consumption: "The global market for marine antifouling coat-
      ings is around 1bn euros, representing 80,000 tonnes of paint per annum."
      InnovateUK.org Webpage: Cleaning up antifouling, April 29, 2010, https://
      connect.innovateuk.org/web/smart-materials/cleaning-up-antifouling;
      Frost & Sullivan's Study on Potential Market for Carbon Nanomaterials'
      Applications, Chart (page 8) of CNT Market Estimation for Coatings (2009-
      2015) shows 2015 revenues projected at US $2.5 billion, Frost & Sullivan,
      February 28, 2011, http://www.nist.gov/cnst/upload/Valenti-NIST.pdf.

175   Biofilms are clumps of: Maria Esperanza Cortés; Jessika Consuegra Bonilla;

Ruben Dario Sinisterra, Biofilm formation, control and novel strategies for eradication, Science against microbial pathogens: communicating current research and technological advances, Formatex 2011, http://www.formatex. info/microbiology3/book/896-905.pdf; NIH Guide: Research on Microbial Biofilms, December 20, 2002, http://grants.nih.gov/grants/guide/pa-files/ PA-03-047.html/.

175   Scientists on a marine expedition: AskNature.org: Biofilm-inhibiting chemical protects surfaces: red alga, http://www.asknature.org/strategy/ b5fc580c4821fb11d9264adcfec126aa; UNEP Press Release: Entrepreneurs of the Natural World Showcase Their Groundbreaking Solutions to the Environmental Challenges of the 21st Century, May 28, 2008: Ibid.

176   In 2010, BioSignal: Bloomberg Businessweek Webpage: Company Overview of Biosignal Ltd., http://investing.businessweek.com/research/stocks/private/ snapshot.asp?privcapId=704876.

176   Now an Arizona inventor: Monarch Power Press Release: Arizona Innovator Raises Solar Lotus Through Biomimicry, February 29, 2012, http:// www.businesswire.com/news/home/20120229005298/en/Arizona-Innovator-Raises-Solar-Lotus-Biomimicry.

176   More than two thousand years ago: Kishore Balakrishnan, "Padmapatramivambhasa," February 25, 2005, http://www.carnatic.com/kishore/ blog/2005/02/25/padmapatramivambhasa/.

177   German scientist Dr. Wilhelm Barthlott: Lotus Effect Webpage: Professor Dr. Wilhelm Barthlott, http://www.lotus-effekt.de/en/uns/details/barthlott .php;

177   More than three hundred thousand buildings: Lotus Effect Webpage: Frequently asked questions, http://www.lotus-effekt.de/en/faq/index.php.

178   Researchers at Shanghai Jiao Tong: Zhou, H., Li, X., Fan, T., Osterloh, F. E., Ding, J., Sabio, E. M., Zhang, D. and Guo, Q. (2010), Artificial Inorganic Leafs for Efficient Photochemical Hydrogen Production Inspired by Natural Photosynthesis, Adv. Mater., 22: 951–956. doi: 10.1002/adma.200902039

179   Dyesol, a company based in Australia: DyeSol company Website: http:// www.dyesol.com/.

180   Daniel Nocera: Alexis Madrigal, Cheap Catalyst Could Turn Sunlight, Water Into Fuel, Wired Science, July 31, 2008, http://www.wired.com/ wiredscience/2008/07/reverse-fuel-ce/; Christian Science Monitor, MIT Researchers Attain Solar Nirvana, August 1, 2008, http://www.csmonitor .com/Environment/Bright-Green/2008/0801/mit-researchers-attain-solar-nirvana.

180   The U.S Fish and Wildlife Service: American Bird Conservancy Press Release: Bird Deaths from Wind Farms to Continue Under New Federal Voluntary

Industry Guidelines, February 8, 2011, http://www.abcbirds.org/newsandreports/releases/110208.html; http://www.abcbirds.org/abcprograms/policy/collisions/pdf/Bird_mortality_estimate_032212.pdf.

180 Instead, inspired by the way: Anne Ju, Students harness vibrations from wind for electricity, Cornell University Chronicle Online, May 25, 2010, http://www.news.cornell.edu/stories/may10/vibrowind.html.

181 By studying the samara: Jacob Aron, Spinning seeds inspire single-bladed helicopters January 28, 2011, *New Scientist* Webpage: http://www.newscientist.com/article/dn20045-spinning-seeds-inspire-singlebladed-helicopters.html.

181 The maple samara has inspired: Coroflot company Webpage: Edward Bae Samara, http://www.coroflot.com/edwardbae/homefurniture.

182 Another plant, from one of the hotter: UNEP Press Release: Entrepreneurs of the Natural World Showcase Their Groundbreaking Solutions to the Environmental Challenges of the 21st Century, May 28, 2008: Ibid.

184 More than 50 percent: Global Deforestation, University of Michigan, January 4, 2010, http://www.globalchange.umich.edu/globalchange2/current/lectures/deforest/deforest.html; Facts about rainforests, The Nature Conservancy, October 6, 2011, http://www.nature.org/ourinitiatives/urgentissues/rainforests/rainforests-facts.xml

185 With continued extinctions: E.O. Wilson, "The current state of biological diversity." In *Biodiversity*, ed. E.O. Wilson, (Washington, D.C.: National Academy Press, 1988) 18; Encyclopedia of Life Webpage: Magnoliophyta — Flowering Plants, http://eol.org/pages/282/overview.

185 Researchers have confirmed: Lee Eng Lock, in discussion with the author, June 2000.

185 Load-bearing or protective structures: AskNature.org: Lightweighting: Scots pine, http://www.asknature.org/search?category=default&query=scots+pine+lightweighting.

185 As biomechanics researcher and author: AskNature.org: Leaves resist crosswise tearing: grasses, http://www.asknature.org/strategy/bf49fa9c75f567967bc5fc1b24d081e7.

186 Santiago Calatrava showed: Wayne Lorentz, The Chicago Spire, Chicago Architecture Info Webpage: http://www.chicagoarchitecture.info/ShowBuilding/357.php.

187 When architect Layla Shaikley: Ehsaan Mesghali, Layla Shaikley's Pomegenerate, Biomimetic Architecture Webpage, February 10, 2010, http://www.biomimetic-architecture.com/2010/pomegenerate/

188 There are more than twelve hundred: JMBamboo Webpage: Go green with

bamboo, http://www.jmbamboo.com/2011/04/go-green-with-bamboo/; Allaboutbamboo, http://www.jmbamboo.com/2011/09/all-about-bamboo/.

188   After World War II: David Farrelly, *The Book of Bamboo*, (Sierra Club Press, 1984), 16.

188   Thomas Edison's first: David Farrelly, 5: Ibid.

189   So capable was the junk: Kallie Szczepanski, Zheng He's Treasure Ships, About. com, http://asianhistory.about.com/od/china/p/Zheng-Hes-Treasure-Ships .htm.

189   An archeological dig near Nanjing: Sally K. Church, "Two Ming Dynasty Shipyards in Nanjing and their Infrastructure," Page 3, Chapter 3, Shipwreck ASIA: *Thematic Studies in East Asian Maritime Archaeology*, Edited by Jun Kimura, 2010.

190   During and after World War I: Richard P. Hallion, "Wooden Aircraft and the Great War," Journal of Forest History, October 1978, 200-202, http:// www.foresthistory.org/Publications/JofFh/Hallion.pdf.

190   The annual global market for glass fiber: World Glass Fiber Reinforcements Market Overview by Industry Experts, PR Web, May 23, 2012, http://news.yahoo.com/world-glass-fiber-reinforcements-market-over-view-industry-experts-162616648.html.

190   Researchers Dr. David Hepworth: Cellucomp company Website: http:// www.cellucomp.com/.

CHAPTER 8: WAMPUM

192   In 2001, scientists duct-taped . . .: Neat Stuff, October 21, 2001, Extreme 2001 Webpage: http://www.ceoe.udel.edu/extreme2001/discoveries/neatstuff/ oct21.html.

193   Fully 4 percent: Industrial Minerals Association North America Webpage: Calcium Carbonate, IMA-NA, 2009, http://www.ima-na.org/calcium-carbonate.

194   Shell beads, called wampum: Tara Prindle, Wampum History and Bacground, Native Tech Native American Technology and Art Webpage: http:// www.nativetech.org/wampum/wamphist.htm.

194   Murex sea snails: Jolique Webpage: Purple Passion, May 15, 2001, http:// www.jolique.com/dyes_colorants/purple_passion.htm.

194   Geerat Vermeij: Carol Kaesuk Yoon, Scientist at Work: Geerat Vermeij; Getting The Feel of A Long Ago Arms Race, February 7, 1995, New York Times Science Webpage: http://www.nytimes.com/1995/02/07/science/

scientist-at-work-geerat-vermeij-getting-the-feel-of-a-long-ago-arms-race .html?pagewanted=all&src=pm.

194   One is from India: Carl Ruscoe, An Insight into Collecting Sinistral Shells, August 31, 2003, British Shell Collectors' Club Webpage: http://www .britishshellclub.org/pages/articles/sini/artsini.htm.

195   This undulating, wavelike movement: Evan Ackerman, Omnidirectional Robot Moves Like a Galloping Snail, July 5, 2011, IEEE Spectrum Webpage: http://spectrum.ieee.org/automaton/robotics/industrial-robots/omni-directional-robot-moves-like-a-galloping-snail.

196   Researcher Derk Joester: Joe Palca, Rock-Munching Mollusks A Model For Artificial Bones, January 13, 2011, NPR Morning Edition, http:// www.npr.org/2011/01/13/132859853/rock-munching-mollusks-a-model-for-artificial-bones.

196   While on the subject of toughness: R. Nederlof and M. Muller, A biomechanical model of rock drilling in the piddock *Barnea candida* (Bivalvia; Mollusca) *J. R. Soc. Interface rsif20120329,* published ahead of print June 13, 2012, doi:10.1098/rsif.2012.0329 1742-5662.

197   The $1 trillion mining industry: Mine: The growing disconnect, PricewaterhouseCoopers LLP, 2012, 1, http://download.pwc.com/gx/mining/ pwc-mine-2012.pdf.

197   Researchers P.U.P.A Gilbert: Gilbert Research Group Webpage: Sea Urchin Teeth, http://home.physics.wisc.edu/gilbert/.

197   Researchers Helmut Colfen and Jong Seto: Jason Palmer, Sea urchin spine structure inspires idea for concrete, February 14, 2012, BBC News Science & Environment, http://www.bbc.co.uk/news/science-environment-17027941.

198   Dr. George Weinstock: Baylor College of Medicine Press Release: Decoded sea urchin genome shows surprising relationship to man, Ross Tomlin, November 9, 2006, http://www.eurekalert.org/pub_releases/2006-11/bcom-dsu110906.php.

199   Brent Constantz: In conversation with the author, November 16, 2011.

202   In a similar way, Calera: Calera company Website: http://calera.com/; David Biello, Cement from CO2: A Concrete Cure for Global Warming? August 7, 2008, Scientific American, http://www.scientificamerican.com/article.cfm? id=cement-from-carbon-dioxide.

205   Kollodis BioSciences of Massachusetts: Kollodis BioSciences company Website: http://www.kollodis.com/.

206   Phillip Messersmith: Messersmith Research Group Mussel Adhesive Protein Mimetics Webpage: http://biomaterials.bme.northwestern.edu/mussel.asp.

206   99 million cubic meters: International Forest Industries Webpage: Global

Demand for Plywood and OSB Increasing, April 27, 2011, http://www
.internationalforestindustries.com/2011/04/27/global-demand-for-
plywood-and-osb-increasing/.

206 Columbia Forest Products: Columbia Forest Products PureBond Brochure,
2006, http://columbiaforestproducts.com/Content/Documents/PureBond_
Brochure.pdf; PureBond Webpage: http://columbiaforestproducts.com/
mobile/PureBond.aspx#/mobile/PureBond.aspx.

207 Our deep ocean shell: Anne Trafton, Iron-plated snail could inspire new
armor, MIT News Office, January 27, 2010, http://web.mit.edu/newsof-
fice/2010/snail-shell.html.

208 If you're not thrilled at the prospect: Giancarlo Paganelli, Cone Shell Webpage: http://www.coneshell.net/Pages/pa_cones_venom.htm.

209 A particularly potent species: Klotz, U., Ziconotide - a novel neuron-specific
calcium channel blocker for the intrathecal treatment of severe chronic
pain - a short review, Int J Clin Pharmacol Ther, 2006 Oct;44(10):478-83;
Baby Joseph, Sheeja S Rajan, Jeevitha.M.V, Ajisha.S.U, Jini.D, Conotoxins:
A potential natural theraputic for pain relief, International Journal of Pharmacy and Pharmaceutical Sciences, Int J Pharm Pharm Sci, Vol 3, Suppl 2,
2011, 1-5.

209 By "blocking the calcium channels: ANI Correspondent, Sea snail
venom paves way for potent new painkiller, DailyIndia ANI, July 3, 2007,
http://www.chninternational.com/cone_snail_venom_attacking_pain
.htm#Sea_snail_venom_paves_way_for_potent_new_painkiller.

209 Nudibranchs, a remarkable and shell-less: Australian Museum Online Sea
Slug Forum: http://www.seaslugforum.net/.

210 As Nan Criqui of the University of California: Nan Criqui, Online E-Mag-
azine Voyager for Kids, Scripps Institution of Oceanography, University
of California, San Diego, http://www.sio.ucsd.edu/voyager/nudibranch/
index.html;   http://aquarium.ucsd.edu/Education/Learning_Resources/
Voyager_for_Kids/nudibranch/nudi4.html; Faulkner, D. J.; Ghiselin, M. T.
(1983), "Chemical defense and evolutionary ecology of dorid nudibranchs
and some other opisthobranch gastropods," *Marine Ecology-Progress Series*
13: 295–301, http://www.int-res.com/articles/meps/13/m013p295.pdf; Ami
Schlesinger & Esti Kramarsky-Winter & Yossi Loya, Active Nematocyst Iso-
lation Via Nudibranchs, Mar Biotechnol (2009) 11:441–444, http://www.tau.
ac.il/lifesci/departments/zoology/members/loya/documents/2009194Ami
SchlesingerMarBiotechActiveNematocystIsolationViaNudibranchs.pdf.

211 Octopi, squid, nautilus: University of California Museum of Paleontology
Webpage: The Cephalopoda: http://www.ucmp.berkeley.edu/taxa/inverts/
mollusca/cephalopoda.php/.

211   Dr. T.H. Waterman: Talbot H. Waterman, A Light Polarization Analyzer in the Compound Eye of Limulus, Science 10 March 1950: Vol. 111 no. 2880, 252-254 DOI: 10.1126/science.111.2880.252.

CHAPTER 9: THE CORPORATE JUNGLE

227   Statoil Corporation: Valeria Criscione, Rush for the barents sea, Nortrade .com, June 26, 2012, http://www.nortrade.com/sectors/articles/rush-for-the-barents-sea/.

238   about $1.2 billion of domestic fans: S Ventilation Equipment Market Set to Return to Growth, Business Wire, June 27, 2011, http://www.businesswire. com/news/home/20110627005705/en/Ventilation-Equipment-Market-Set-Return-Growth-companiesandmarkets.com.

239   As a global-warming gas: U.S. Environmental Protection Agency Web page: Methane, http://epa.gov/methane/scientific.html.

245   In Venice, patents were being granted: Randy Alfred, March 19, 1474: Venice Enacts a Patently Original Idea, *Wired,* March 19, 2008.

245   Now, about 185: World Intellectual Property Organization Webpage: Member States, http://www.wipo.int/members/en/.

CHAPTER 10: DOLLAR SIGNS

249   the owners of dogs have: Wilson CC (1991), "The pet as an anxiolytic intervention," *The Journal of Nervous and Mental Disease* 179 (8): 482–9. doi:10.1097/00005053-199108000-00006. PMID 1856711.

249   Since the eighteenth century, dogs: Kruger, K.A. & Serpell, J.A., "Animal-assisted interventions in mental health: Definitions and theoretical foundations," In Fine, A.H. (Ed.), *Handbook on animal-assisted therapy: Theoretical foundations and guidelines for practice,* (Amsterdam: Elsevier/Academic Press, 2006) 21–38. ISBN 0-12-369484-1.

249   ADHD and autistic kids: Katcher, A.H., Wilkins, G.G.,"The Centaur's Lessons: Therapeutic education through care of animals and nature study," In Fine, A. H. (Ed.), *Handbook on animal-assisted therapy: Theoretical foundations and guidelines for practice,* (Amsterdam: Elsevier/Academic Press, 2006) 153–77, ISBN 0-12-369484-1.

255   There are around two million new businesses: "What's more, the primary source of investment in the 2-2½ million new businesses started each year (which are responsible for almost all net new jobs created each year) is per-

sonal savings and the savings of friends, family and associates . . ." Founder's View: The Myth of the 1%, Corporation for Enterprise Development Newsletter: March 2012, http://cfed.org/newsroom/newsletters/march_2012/. National Venture Capital Association Webpage: Frequently Asked Questions About Venture Capital, http://www.nvca.org/index.php?Itemid=147& id=119&option=com_content&view=article.

260   In fact, in 1985: Thomas Grande, The False Claims Act: A Consumer's Tool to Combat Fraud Against the Government, Health Administration Responsibility Project Webpage: http://www.harp.org/grande.htm.

261   Chinese government loans to solar: Eric Wesoff, Chart of the Day: Solyndra Edition, Greentech Media, August 31, 2011, http://www.greentechmedia. com/articles/read/chart-of-the-day-solyndra-edition/.

CHAPTER 11: REORGANIZING (YOUR) BUSINESS

264   total debt three times that: Joe Weisenthal, This Is What The End Of The Global Debt Super-Cycle Looks Like, Business Insider, December 01, 2011, http://articles.businessinsider.com/2011-12-01/markets/30462034 _1_defaults-cycle-super.

267   A world-famous complex in Kalundborg: Indigo Development, The Industrial Symbiosis at Kalundborg, Denmark, June 12, 2003, http://www .indigodev.com/Kal.html.

268   Nicholas Tee Ruiz: Nicholas Tee Ruiz Website: http://madeinforesthills .com/.

268   A crew from the environmental nonprofit: The Plastiki Expedition Website: http://www.theplastiki.com/.

268   The EcoARK: Miniwiz Sustainable Energy Development Ltd's EcoARK Webpage: http://www.miniwiz.com/miniwiz/en/projects/ecoark.

268   Ugandan refugee Derreck Kayongo: Ebonne Ruffins, Recycling hotel soap to save lives, CNN Heroes Webpage, CNN.com June 16, 2011, http://www.cnn .com/2011/US/06/16/cnnheroes.kayongo.hotel.soap/index.html.

271   As the U.S. government's Energy Star program reports: U.S. Department of Energy, Energy Saver Website: http://energy.gov/energysaver/articles/ lighting-choices-save-you-money.

278   Green chemistry is a subset: John Warner, in discussion with the author, December 21, 2011.

283   One of their initiatives: Treepeople Website: http://www.treepeople.org/ functioning-community-forest.

284 Since then, Interface has: Ray Anderson, *Confessions of a Radical Industrialist: Profits, People, Purpose--Doing Business by Respecting the* Earth (McClelland & Stewart, 2009).

285 Their fastest growing line: Kate Rockwood, Biomimicry: Design Inspirations from Nature October 8, 2008, Fast Company, http://www.fastcompany.com/1036526/biomimicry-design-inspirations-nature2008.

285 As a tribute to Ray in the Guardian: John Elkington, "Ray Anderson, sustainable business pioneer, dies aged 77," http://www.guardian.co.uk/sustainable-business/blog/ray-anderson-dies-interface-john-elkington-tribute.

286 As Sam Stier: Sam Stier, in discussion with the author, December 19, 2011.

288 Do you remember: Jay Harman, Designing the Next Golden Age, Bioneers Conference Plenary October 15, 2004.

289 Let's look again: Louie Schwartzberg is a pioneering cinematographer who specializes in breathtaking, slow-motion or time-lapse images of nature's beauty. His work is a doorway to appreciation of the wonders of nature: http://www.movingart.tv/screeningroom/; http://www.ted.com/talks/louie_schwartzberg_nature_beauty_gratitude.html.

289 Combining our human intelligence with optimism: The Intelligent Optimist is an outstanding resource for positive news and information. Through their magazine and online community, they prove that "The world is a better place than you think," http://www.theoptimist.com/.

EPILOGUE

292 It's often reported: Dave Smith, How to Avoid the Passion Trap, Inc. Magazine, May 5, 2011, http://www.inc.com/articles/201105/how-to-avoid-the-passion-trap.html.

Photo Credits

....

Page

iii   Courtesy of the author

11   Courtesy of the author

15   (top) Courtesy of the author; (bottom) Courtesy of the author

23   (left) Herbert A. 'Joe' Pase III, Texas Forest Service, www.Bugwood.org; (right) Photograph by Jim Cane

25   Public domain (Pinacoteca di Brera, Milano, Italy)

27   (top left) Courtesy of the author; (top center) Public domain (Gaius Cornelius); (top right) Courtesy of Pinnacle Armor; (bottom left) © Kwiktor, Dreamstime.com; (bottom right) Public domain (Takahashi)

33   (top) Courtesy of the author; (bottom) Public domain (Marie-Lan Nguyen)

37   Courtesy of the author

40   (top left) © Jin Yamada, Dreamstime.com; (top center) © Kentannenbaum, Dreamstime.com; (top right) © James Steidl, Dreamstime.com; (bottom left) Public domain; (bottom center) Courtesy of the author; (bottom right) Courtesy of the author

41   (top left) © Kimberley Penney; (top right) © Kimberley Penney; (below) Courtesy of the author

42   (left) © Stephan Pietzko, Dreamstime.com; (right) Courtesy of the author

43   (left) Courtesy PAX Scientific, Inc.; (right) © Dr. Jeremy Burgess/Photo Researchers, Inc.

44   (top left) Courtesy of the author; (top right) © Anatomical Travelogue/ Photo Researchers, Inc.; (bottom left) © Ken Eward; (bottom right) Courtesy of the author

47   (left) © Andrey Korsakov; (right) © Margie Hurwiche, Dreamstime.com

48   Courtesy PAX Scientific, Inc.

49   (left) © Dan Norris, Ancient Images; (right) © Cosmopol, Dreamstime.com

50   (row 1, left) © Michael Elliott, Dreamstime.com; (row 1, right) Public domain (Dartmouth College); (row 2, left) Public domain (NASA); (row 2, right) Mamahoohooba, Dreamstime.com; (row 3, left) Arlenenorton, Dreamstime.com; (row 3, right) Courtesy of the author; (row 4, left) Public domain (NASA); (row 4, right) © CERN

51   (left) Public domain (Chambers' Encyclopedia, 1875); (right) Courtesy of the author

53   © Rachael M Bertone

54   (left) Courtesy of the author; (right) Courtesy of the author

55   (top left and right) Courtesy of the author; (below) Courtesy of the author; (bottom) Courtesy of the author

60   Courtesy of the author

83   Public domain (Pennie Gibson)

87   © Sharklet Technologies

90   © Sharklet Technologies

94   © Speedo

96   Courtesy of BioPower Systems Pty Ltd

100   (left) © Ronald Adcock, Dreamstime.com; (right) © Joe Subirana, WhalePower Corporation

102   Public domain (Meyers Konversationslexikon from 1885–90)

103   Public domain (NOAA)

108   (left) Public domain (NASA); (right) Courtesy of the author

109   Courtesy of the author

119   © Isselee, Dreamstime.com

121   (left) Public domain (David Clements); (right) © Dr. Jeffrey Karp

126   Public domain (Dumplestilskin)

128   (left) Public domain (Alok Mishra); (right) Public domain (C. Schlawe, U.S. Fish and Wildlife Service)

129   Courtesy of the author

132   © Ozflash, Dreamstime.com

136   © Festo

142   © Art Gray, Panelite ClearShadeTM IGU photographed at the Henry Madden Library, California State University, Fresno (www.panelite.us)

Index

A. O. Smith Corporation, 242–244
acupuncture, 43
adhesives, 7, 116, 121, 139, 205–206
Adventure Ecology, 268
aerodynamics, 92, 93, 134, 139
Africa, 150, 159
air-conditioning, 61, 62, 105, 120, 150,
    224, 225, 267
aircraft, 80; bamboo used in, 188;
    construction of, 190; deicing, 156
    fluid dynamics of, 88–89;
    maneuverability of, 135, 139–140;
    nature-based design of, 128, 134;
    stalls by, 98; wing design on, 133, 144
albatross, 127, 134
Alcoa, 182
Alzheimer's disease, 3, 209
Amazon, 273
Anastas, Paul T., 279
Anderson, Ray, 284–285
angel investors, 253–254, 258
anhydrobiosis, 161
antibiotics, 3, 19, 31, 119, 138, 141, 149,
    167, 176
anticoagulants, 113, 114–115
antiviral treatments, 167
ants, 151–152
Archimedes' screw, 51
architecture
biomimicry in, 21, 25, 73, 151, 186–187;
    honeycomb structures in, 143;

spiral pattern in, 40, 45, 52
armor, 24, 26, 27
art, spiral pattern in, 52
artificial hands, 149
artificial intelligence, 169
artificial muscles, 117
artificial photosynthesis, 178–179
ARUP construction company, 150
asbestos, 234, 235–236
AskNature.org, 70–71, 288
atmospheric mixing, 63
Ausubel, Kenny, 65, 66, 76, 78–79, 227
autism, 3
automobiles, 69, 92, 237. *See also*
    vehicles

Babcock, Jim, 280
bacteria, 31, 149, 162–163, 175
baleen, 103–104
ball bearings, 120
bamboo, 187–189
barbed wire, 27
Barthlott, Wilhelm, 177
Bateman, Bill, 100
bats, 180
batteries, 180, 269
Baumeister, Dayna, 70–73, 74, 229, 286
Bayer Corporation, 30
Bedo, Daniel, 163
beer, 24, 62, 168

bees, 7, 140–145, 154
beeswax, 141–142
beetles, 159–160
bee venom, 114
Benyus, Janine, 4, 18, 70, 71, 266, 267, 272, 276, 284, 286
Bertone, Francesca, 56, 223
Beyond Benign, 282
biodegradable products, 121, 278
biodomes, 162–163
biofilms, 90–92, 162, 175–176
biofuels, 268
bioleaching, 173
bioluminescence, 106, 107, 117
Biomimetics Innovation Center, 174
biomimicry, 88, 217; business start-ups in (*see* entrepreneurialism); as cross disciplinary, 216–217; defined, 2, 18; energy efficiency of, 7–8, 20, 73; form based, 88; and green chemistry, 278, 280; as investment, 65, 68, 259; markets for, 70; motivation for, 217; need for, 2–4, 66, 291–293; origins of, 4–5, 18; platform technologies in, 176; potential of, 6–7, 20–23, 31, 80–81, 250; product development in, 217–218; public-private partnerships in, 67; resources for, 70–71, 212; value generation via, 73–74
Biomimicry (Benyus), 4, 266
Biomimicry BRIDGE, 67
Biomimicry Design Challenge, 117
Biomimicry Guild, 70, 73
Biomimicry Institute, 70, 71
Biomimicry 3.8, 70, 71, 72, 229, 286–288
biomineralization, 200
Bioneers, 65–66, 76, 78, 227, 269
BioPower Systems, 94–96
BioSignal Ltd., 175–176, 236
biosphere, 275–276
Biostability Ltd., 181
bioSTREAM project, 95
biotherapy, 114, 116, 138
birds, 128–136, 146, 155, 180
blackflies, 163
blood, 114, 115, 119
blowflies, 138–139
boats. *See also* ships; design of, 24, 108, 215–216; fluid dynamics for, 43, 56;

hydrodynamic films for, 134; natural materials for, 188–189; stress-resistant fabrics in, 186
Boeing, 128, 133, 134, 190
bones, 196, 200–201. *See also* skeletal system
Brennan, Anthony, 90, 91, 112
Brinker Technology, 115
buildings, 177, 187, 187, 271
business, 67–70; failures of, 292–293 free market impact on, 76; low-risk entry into, 190, 241; sustainable life cycle of, 264–277; transferred from research labs, 92, 100–101, 250; value chain in, 57, 237
business model, 241, 247; biomimicry effect on, 4; environmental considerations in, 21; for green chemistry company, 280; for international growth, 204; licensing in, 101; for mail order, 170–171; of PAX, 59–60, 252; and profitability, 247–248
butterflies, 20, 72, 153–156
Byetta (exenatide), 119

Calatrava, Santiago, 186
calcium carbonate, 193, 198
Calera, 199, 202–204, 205, 229, 286
California Energy Commission, 261
camouflage, 28
cancer, 3, 114, 170, 235, 249–250, 279
Cannon, Amy, 282
carbon dioxide, 45, 62, 89, 202–204, 286, 288
carbon footprint, 67, 274
cardiovascular system, 200
Carrier, 224–226
CelluComp, 190
cement, 199–204, 286
Center for Biology-Inspired Design (CBID), 287–288
Center for Interdisciplinary Bio-inspiration in Education & Research (CiBER), 287
Center for Turbulence Research (Stanford University), 187, 220
Centre for Bioinspiration, 67

cephalopods, 211
cetaceans, 98, 102, 108
C40 Climate Leadership Group, 67
chameleons, 122
Charles, Prince of Wales, 67
chemical manufacturing, 21, 22,
    278–282, 286
Chicago Spire, 186
China, 63; bamboo use in, 188–189;
    biomimicry applications in, 73;
    chemical research in, 281; expanding
    economy in, 204–205; loans to solar
    companies in, 261; patent law in, 245;
    sustainability activity in, 65, 66;
    venture capital in, 255; wind power
    in, 101
chiton, 196
ChromaFlair pigment, 155
Chu, Steven, 66
Ciamillo, Ted, 111
clean technology, 21, 61, 65, 66, 67, 259
ClearShade glass, 142–143
climate change, 63, 67, 78
cockroaches, 147–150
coconuts, 174
codeine, 30
Cölfen, Helmut, 197
collision protection, 129, 130, 139–140,
    144, 146, 185
coloration, 72, 154, 155
Columbia Forest Products, 206
communication systems, 109–110,
    276–277
compact fluorescent lightbulbs (CFLs),
    271
concrete, 202–203
Cone, Clarence, 135
Confessions of a Radical Industrialist
    (Anderson), 285
conjugated linoleic acid (CLA), 240
Constantz, Brent, 199–205, 229, 250
construction industry, 21, 65
consumers, 237, 275; buying decisions
    by, 234, 246; dealing directly with,
    241–242; demand for green products,
    68;
cooperation, as survival imperative, 269
coral, 193, 200, 202, 288
Córazon Technologies, 200

Coumadin (warfarin), 115
Cox, Joseph, 28
crabs, 42, 198–199
Criqui, Nan, 210
Crohn's disease, 116
CSIRO, 145, 241
Curran, 190

Dacke, Marie, 145
Dampier Archipelago, 85
Da Vinci Index, 21–23
Debray, Alexis, 122
deforestation, 182–184, 188, 192, 206,
    240
Delphi auto parts, 226–227, 242
de Mestral, George, 7
dengue fever, 158
Denmark, 66
dermal denticles, 87–92
desertification, 240, 241
DeVoe, Irving, 162
Dewar, Stephen, 100–101, 112
diabetes, 3, 119, 139, 158
DiagNose RASCargO, 250
Diamond, Gill, 119
diamonds, 46, 47
Dickinson, Michael, 139
digoxin, 30
diseases, 114, 116, 158–159, 161, 249–250
disruptive pattern material (DPM), 28
DNA, 44, 164, 198
dogs, 114, 249–250
Dollar, Aaron M., 149
dolphins, 103, 106–111
Dragon Skin, 26, 27
durian fruit, 187
Dutch East Indies Company, 14–16
Dyesol, 179

echinoderms, 197, 198
Eckels, Steve, 62
EcoARK, 268
The Ecology of Commerce (Hawken), 284
economies; developed vs. Third World,
    204; local vs. global, 277; and natural
    resource supply, 275; and new
    technology investment, 65, 231;

and venture capital, 255; wealth concentration in, 76

ecosystems, 70, 74, 265, 274, 276; in business, 269–270; in neighborhoods, 283

Edward Bae Designs, 181

efficiency; of biomimicry-inspired designs, 7–8, 20; in buildings, 144; of eggs, 192–193; in electronic screens, 154; in flight, 133–134; of human athletes, 93; of natural systems, 37, 46, 87; of network pathways, 169, 185; of passive ventilation designs, 150–151; of propellers, 181; in rotor design, 53; and survival, 270

Eggleton, Paul, 150

eggs, 191–193

Electric Boat Company (EBC), 222

electricity, in air-conditioning, 150–151; generation of, 94–96; solar-generated, 178–180; use by water pumps, 185

Eli Lilly, 119

energy. *See also* fossil fuels; access to, 288–289; business use of, 58, 270–272; hydroelectric, 94–96; for lighting, 156; natural movement of, 52; to overcome drag, 45–46, 87–88, 89, 92, 135, 174; solar, 178–180, 278; in turbulence, 64, 107; wasteful use of, 3, 5–6, 52–53, 59, 234

Energy Research Group (ERG) Australia, Ltd., 35–36

Energy Star program, 69, 234, 243, 271

engineering, 130; biomimicry impact on, 21, 22, 220–221; paradigm shifts in, 229; vs. science, 236; termite models in, 151

entrepreneurialism. *See also* business; in biomimicry vs. traditional business, 216–218; choosing markets for, 204–205; funding for, 200–201, 203, 251–261, 281, 293; in mail-order business, 170–172; market cultivation in, 234–248; motivation in, 262–263, 282; natural resources for, 111–112; and small companies, 72–73, 92–93; teamwork in, 229–233, 270; technology role in, 218–229; vision in, 100, 230, 232

Envira-North Systems Ltd., 101

environment. *See also* climate change; global warming; pollution vs. business interests, 20–21, 67–70; clear-cutting effect on, 182, 192; disasters, 2–3, 6, 18, 291–293; and energy efficiency, 52, 89, 288–289; man-made chemicals in, 278; proactive approach to, 73–74, 274

enzymes, 162, 173

ephedra, 29, 30

epilepsy, 114, 250

ergot, 30

Evans, Bob, 111

EvoLogics, 110

evolution, 46, 49, 51, 291; of birds, 133; of butterflies, 156; of cockroaches, 149; of flies, 157; of humans, 5; of mollusks, 194; of mycelia, 166, 168; of natural armor, 26; in nature, 3, 4, 88; of sea urchins, 198; of seeds, 181 of sharks, 86, 87; of snakes, 124; of superbugs, 31, 167; of survival strategies, 18, 74, 78, 87; of whales, 102–105

Eward, Kenneth, 44

exenatide, 119

extinction, 19, 184–185

eyes, 7, 143–146, 211

Fan, Yubo, 129

fans; as business venture, 101, 238–239 design of, 8, 54–56, 59, 99, 181, 244 energy use by, 101–102, 226

Fastskin (Speedo), 93

Fastskinz MPG-Plus wrap, 92

feathers, 127, 128, 135, 154, 155

Fermanian Business & Economic Institute (FBEI), 21

Festo, 135–136

fibers, 145-146, 155, 165, 189-190, 235, synthetic, 189–190

Fibonacci numbers, 176

filtration technology, 104, 124

Fish, Frank, 97–100, 112

Flammang, Brooke, 95

fleas, 145

Fleischmann, Wilhelm, 138

Fleming, Alexander, 31
flight. *See also* aircraft; of bees, 144–145; of birds, 127–128, 133–136; of flies, 139;
flounder, 27, 28
fluid dynamics, 26, 43, 45–46, 49, 51–56, 88, 110
food products, 21, 62
ForceFins, 111
forests; as business models, 264–266, 267; clear-cutting of, 182–184, 188, 192, 206, 240; urban, 283
formaldehyde, 206
fossil fuels, 45, 52–53, 112, 128, 174, 178, 202, 227–228
fossils, 124, 193
fractals, 42
Fraunhofer Institute for Manufacturing Engineering and Applied Materials Research, 88
frigate birds, 130
Full, Robert J., 149
Fuller, Buckminster, 44
fungi, 164–172, 182
Fungi Perfecti, 170
furanones, 175

gastropods, 211
Gates, Robert, 66
Gaudi, Antoni, 187
GE, 68, 72
geckos, 7, 120–122
General Dynamics, 222, 260
genetically modified organisms (GMOs), 162
geometry, 24, 25, 40, 42, 52; in liquid flow, 46
Gibson, Lorna, 129
Gielda, Tom, 62
Gila monster, 119
Gilbert, P.U.P.A., 197
Glaswerke Arnold, 146
Glidden, Joseph, 27
global positioning systems (GPS), 145
global warming, 6, 62, 239
glucometers, 139
glue. *See* adhesives
Goeze, Johann, 161

golden ratio, 41
gold mining, 162, 172–173
governments, 69, 279; academic research funded by, 281; energy incentive from, 270; grants from, 259–262; and proprietary technology, 224, 245; protectionist policies of, 238
grants, 22, 259–262
grass, 185–186
Grassberger, Martin, 138
Grätzel, Michael, 179
Great Recession, 65, 76, 255, 293
green chemistry, 8, 22, 278–282, 286
Green Chemistry (Anastas and Warner), 279
GreenerChoices.org, 69
greenhouse gases, 3, 234, 239–241, 271, 275–276, 286
greenwashing, 68–69
gross domestic product (GDP), 20, 264
groundwater contamination, 172

hands, artificial, 149
"Harmony: A New Way of Looking at Our World" (film), 67
Harvard University, 95, 149, 156–157
Hawken, Paul, 58–59, 60, 227, 244, 267, 284
health care costs, 105–106
heart disease, 115
heart pacemakers, 105–106, 116
heat detection, 125–126, 134, 157
heat exchangers, 105, 150
Helbing, Dirk, 151
helicopters, 181
Hepworth, David, 190
heroin, 30
Heyning, John, 105
hippopotamus, 1–2, 103
HOK architectural firm, 73
Holbein, Bruce, 162
Honda, 68
honey, 141
honeycomb, 142–143
Howe, Laurens, 100
Howe, Robert D., 149
Hui, Joseph, 176
human beings; cardiovascular system, 6;

as disconnected from nature, 71–72, 289; energy efficiency of, 46, 288; immune system of, 167, 170; insatiable nature of, 74–77; intelligence of, 292; populations of, 19, 275; reproduction in, 198; resistance to change, 78–81; spiral patterns in biology of, 43; survival of, 19, 292
hummingbirds, 127
Huntington, Collis, 78
hurricanes, 45, 62, 206
HVAC (heating, ventilation, and air-conditioning), 61, 144, 150–151
hydrocarbons, 168
hydrodynamics, 98, 134
hydroelectric power, 94–96
hydrogen fuel, 178–180

ice, 156–157
imaging technology, 126
impellers, 94–96
India, 204, 255, 281
industrial chemistry, 278–282, 286
industrial ecology, 266, 267
industrial revolution, 3, 5, 45, 52, 67–68, 292
industry, 3, 247, 276
infections, 175–176
intellectual property, 31, 58, 232, 245, 252
intelligence; artificial, 169; human network of, 292; in invertebrates, 211
Interface, Inc., 72, 284–285
internal combustion engine, 192
investors. *See also* venture capital; angel, 58, 253–254, 258; evaluating, 233; incentives for, 245
Isaacs, Andrew, 247
Izola, Slovenia, 143

Japan, 109, 204, 265
JDS Uniphase Corporation, 155
Joester, Derk, 196
Johnson, Ray, 226, 227
Johnson & Johnson, 200
Joint Center for Artificial Photosynthesis, 178

Joint Improvised Explosive Device Defeat Organization, 250

kangaroos, 240–241
Kavehrad, Mohsen, 109–110, 112
Kayongo, Derreck, 268–269
Kelly, Michael, 27
keratin, 104, 120
Khosla, Vinod, 61, 203
Killian, Christian, 197
kinetic energy, 145, 180
Kochian, Leon, 173
Kogyo, Tanaka Kinzoku, 155
Kollodis BioSciences, 205
Komodo dragons, 119

laser range finder (LRF), 143–144
leaves, 178–180
Lee, Simon, 149
leeches, 113–116
Lekoudis, Spiro, 222
Leonardo da Vinci, 26
Li, Kaichang, 206
licensing, 58, 101, 242
lightbulbs, 271
light-emitting diodes (LEDs), 156
Lilley, Geoffrey, 135
limpets, 196, 207
Lipkis, Andy, 283
Liquid Life Lab, 98
lizards, 119–120
locomotion, 149–150, 195, 209
Lotusan, 177
lotus plants, 176–177
Lovins, Amory, 224, 261
Lunocet, 111

MAD Architects, 143
maggots, 136, 137
magnetic fields, 43, 120
magnetite, 196, 207
malaria, 30, 158
Malheur National Forest (Oregon), 164
Mandelbrot, Benoit, 42
markets, 76; for biomimetic products, 70, 218; choosing, 204–205, 232, 241;

cultivating, 234–248; for green
    products, 68–70
Mars, 187
Massachusetts Institute of Technology
    (MIT), 95, 121, 129, 180, 207
materials
    inherent properties of, 280
    sustainable use of, 273–274, 276
mathematical ratios, 25, 40, 44, 176
Max Planck Institute (Germany), 121,
    139, 146
McEwan, Ian, 115
McLaughlan, Neil, 140
Mead, James, 105
Messersmith, Phillip, 206
metals; reclaiming, 162, 268; toxic waste
    from, 168, 172–173
methane, 239, 240–241
Mid-Course Correction (Anderson), 285
mining, 162, 182, 197, 203
Mirasol screen technology, 154
Mohs' scale, 196
molds, 165, 168, 177
mollusks, 42, 193–197, 207–211
morphine, 30
Morphotex, 155
mosquitoes, 157–159
movement. *See also* locomotion;
    efficient vehicular, 151–152;
    spiral paths of, 47–49, 51
Muiron Islands, 122
muscles, 117
mushrooms, 43, 164–172
mussels, 205–206
mycelia, 164–172

Nakatsu, Eiji, 128, 130
Namib Desert, 159–160
nanofibers, 190
nanoparticles, 88, 179
nanotechnology, 8, 194, 197
NASA Langley Research Center, 133–135
National Institutes of Health (NIH), 92,
    170
National Science Foundation (NSF), 121,
    207, 287
Native Americans, 186, 194
Natura company (Brazil), 72

nature; as design model, 3–6, 18, 20, 56,
    74, 128, 187, 220, 289; efficiency of,
    7–8, 37, 46, 51, 87, 237; human
    disconnect from, 71–72; inter-
    dependence in, 269; movement
    patterns in, 47–49, 52, 64;
    optimization processes in, 272–273;
    as population equalizer, 19; problem
    solving in, 192; as resource, 111, 212,
    275, 291; vs. science, 225
nautilus, 33, 40, 56, 211
Neal, Murray, 26
Netherlands, 66
networks, 165, 168–169, 185
neurological system, 146, 168, 169
neutrinos, 80
nicotine, 234–235
Nissan Motor Company, 143–144, 155
Nocera, Daniel, 180
noise reduction, 127, 128, 135
Norian Corporation, 200, 201
noses, 250
nuclear waste, 192
nudibranchs, 209
nutritional labels, 69–70, 282

Oakey, David, 285
oceans, 95-96; acid buildup in, 288;
    waste accumulation in, 290–291
open source model, 245
operculum, 207–208
opium, 30
optics technology, 155, 211
oriented-strand board (OSB), 206
Ornilux glass, 146, 147
Orr, David, 261
Ortiz, Christine, 207
Øverli, Bente, 69
owls, 127, 128, 135
Oxford Biomaterials Ltd., 146
Oxford Silk Group, 146

pain control, 209, 210
paint, 88–89, 155, 174, 177
Pak, Kitae, 159
Panelite glass, 142–143
paper, 188, 192, 273

Pareto principle (80–20 rule), 271
Park, Sungmin, 129
partnerships, 67, 227, 232, 252–253, 259
patents, 22, 56–58, 171, 177, 244–246,
    251, 282
PaxFan, 8, 59, 102, 181, 226–227,
    242–244, 261
PAX Mixer, 61, 63
PAX Scientific, Inc., 3, 23, 51, 54, 55, 70,
    228, 230, 259–260, 261 276–277;
    business strategy of, 56–64, 219, 220,
    242–244, 247, 252; investors in,
    253–256; markets for, 232, 237, 238;
    product development at, 221–226,
    236, 242; sustainability effort at, 266;
    water stagnation prevention, 159
PAX Streamline, 61–62, 256
PAX Water Technologies, Inc., 60, 61
Pearce, Mick, 150
penicillin, 167
Pennsylvania State University, 109, 133,
    223
Peschka, Manfred, 88, 89
pharmaceuticals, 29–31, 167–168
photosynthesis, 176, 178–180
phytoremediation, 173
piddock, 196–197
Pinnacle Armor, 26
placoid scales, 87
plants, 29–30, 176, 185
plastics, 21, 192
plywood, 206
polarization, 211
polio, 161
pollination, 154, 157
pollution, 6, 63, 274; in balanced
    biosphere, 275–276; from cement
    production, 202–204; fungi to
    remediate, 168, 172–173; from
    internal combustion engine, 192;
    from ships, 174
Polynesia, 75–76
poo-gloo, 162–163
porpoises, 103, 108–111
possums, 117–118
pressure waves, 208
Prialt (Ziconitide), 209
Procter & Gamble, 73
product development, 217, 218, 221, 274

profitability, 171, 229. 201; and
    anticipating trends, 31; of biomimicry
    applications, 68, 73, 74, 136, 218, 231;
    and business model, 247–248; and
    sustainability, 20–21, 285
propellers, 54–56, 88, 95, 107, 181,
    221–225
proportions, 24–26, 40, 41
protective structures, 185–187, 192–193,
    199
prototyping, 54, 57–58, 133, 220, 221,
    224, 251
public-private partnerships, 67
pumps, 54–56, 62, 185
Purebond, 206

Qualcomm, 20, 72, 154
quinine, 30

radulas, 195–196
rain forests, 184, 240, 265
rainwater management, 73, 283–284
ratites, 132–133
Reaser, Lynn, 21
Rechenberg, Ingo, 120
recycling, 8, 168, 267–269, 273, 276
refrigeration, 59, 62–63, 65, 105, 161,
    224, 243
REGEN Energy, 144
research and development (R & D), 57,
    101, 243, 280–281, 287
resilin, 145
return on investment (ROI), 247–248,
    255, 259
Reynolds, Jorge, 106
Riddell, Richard, 222
Riddell helmets, 129
risk avoidance, 227, 228
roadways; congestion on, 151–152;
    deicing, 156–157; dirt reduction on,
    177; as networks, 169
robots, 122, 144, 145, 150, 195, 287
Rocky Mountain Institute, 224
roofing materials, 235–236
Roser, Bruce, 181
Ross, Sir Ronald, 158
rotor design, 53–56

rubber, 145
Ruiz, Nicholas Tee, 268

saguaro cactus, 187
Sahara sandfish, 120
sales, 20–23, 242
salt, 62, 114
San Diego Zoo, 21, 67
saw chains, 28
science, 52, 79-80, 225, 236, 289
seashells, 37, 39, 40, 44, 193–197, 200,
    208–209
sea snakes, 122–124
sea urchins, 197–198
seaweed, 32, 37, 74, 137, 174, 175, 187
seeds, 181, 183
self-cleaning mechanisms, 20, 104, 121,
    156, 176–177
self-similarity, 42
Seto, Jong, 197
Shadyac, Tom, 79
Shaikley, Layla, 187
Sharklet Technologies, 90–92, 236
sharks, 85–96
shells. *See* eggs; mollusks; seashells
Sherman, Ronald, 138
ships. *See also* boats; biomimetic
    technology applied to, 87–91;
    construction of, 188–189; pollution
    from, 174
shipwrecks, 14–16
shock absorbers, 129
shopping locally, 277
silicon dioxide, 196
Simons, Nina, 66
Sinosteel skyscraper (Tianjin, China),
    143
SinsofGreenwashing.org, 69
Skeletal Kinetics, 201
skeletal system, 146, 193, 200
skin; biomimetic, 92–94; of blowfly
    maggot, 138; dolphin, 108; lizard,
    120; shark, 87–92, 174
SkinzWraps, Inc., 92–93
skyscrapers, 143, 186, 188
slime mold, 164, 169
slugs, 193, 195
smallpox, 161

snails, 41–42, 193, 194, 195, 207, 211
snakes, 121–126
solar energy, 178–180, 278
solar system, 47, 48
Solyndra, 260
Southwest Airlines, 135
species; accelerated disappearance of,
    184; cooperation among, 168;
    proliferation of, 3–4; as resource, 185;
    survival of, 6, 7, 18–19
Speedo, 93
spiders, 145–146
spinal cord damage, 146
spiral pattern; in architecture, 40, 45,
    187; of ears, 208; in egg shape, 193;
    equiangular, 41, 45; in fluid dynamics,
    46, 53–56; as foundation of universe,
    52; in nature, 37, 39, 42–43, 49, 50;
    in paths of movement, 47–49, 51; in
    plants, 176; in religious imagery, 34,
    39–40; technological potential of, 64,
    102, 224
spores, 168, 172
sporting goods manufacture, 190
Srinivasan, Mandyam, 145
Stamets, Paul, 165–172, 250
Stanford University, 187, 199–200, 202,
    220
Statoil Corporation, 227–228
Stenzel, Volkmar, 88
Stephens, Iain, 254
Stier, Sam, 286, 287
Sto Corp., 177
StopGreenWash.org, 69
straight lines, 46–47, 49, 52, 53
stress resistance, 185–186
Stürzl, Wolfgang, 145
superbugs, 31, 149
superhydrophobic effect, 177
Suresh, Subra, 207
surface textures. *See also* skin
    on boat hulls, 134; of butterfly wings,
    156; on manufactured wings,
    133–134; owl feathers, 127, 128;
    water-repellent, 176–177
SustainAbility, 21
sustainability, 8, 78, 282; academic
    support for, 287; biomimetic core of,
    65, 74; and business life cycle,

264–277; vs. greenwashing, 68–69; and industrial revolution, 292; worldwide investment in, 20, 66–67
swarm logic, 144
Swedish Center for Biomimetic Fiber Engineering (Biomime), 287
swimmers, 93, 111, 130
synergism, 168, 171, 270
Synthes Holding AG, 200, 201

tardigrades, 160, 161
tax incentives, 21, 67
teamwork, 229–233, 270
Technical University of Berlin, 120
technology, 24, 26, 192, 291; in biomimicry vs. existing industry, 218–229; clean, 21, 61, 65, 66, 67, 259; nano, 8, 194, 197; platform, 176, 242; research-to-business transfer of, 92, 218; resistance to new, 80, 229, 237; spiral geometry in, 64
teeth, 195–196, 197
Teijin Fibers Limited, 155
termites, 150–151
TerraChoice, 68, 69
textiles, 21, 93–94, 155, 186
thermoreceptors, 125–126
3-D printing, 54
time, as resource, 272
Titanic, 162, 189
Toba catastrophe theory, 19
tornadoes, 41, 62
toxicology, 278–279, 281, 282
trains, 128
transportation, 21, 130, 277
TreePeople, 283–284
trees, 273, 283-284; as building design model, 186; felling, 28, 182–184; water storage in, 283–284
tuna, 95, 111
turbines, 55, 94–96. *See also* wind turbines
turbulence, 46, 53, 64, 107, 135, 181
turtles, 43

Ulrich, Evan, 181
ultraviolet light, 146, 278

Underwriters Laboratories, 68, 69
United Technologies, 225
University of California, Berkeley, 121, 129, 149, 228, 247, 287

vaccines, 161, 181–182
van der Waals forces, 121
vehicles, 92; deicing, 156–157; road configuration for, 151–152; safety of, 130, 139–140, 143–144, 185
Velcro, 2, 7
venture capital (VC), 61, 65, 171, 200–201, 203, 245, 246, 254–259, 281
Vermeij, Geerat, 194
Vincent, Julian, 138–139
Viney, Christopher, 1
Vogel, Steven, 124, 185
Vollrath, Fritz, 146
Vukusic, Peter, 156

wampum, 194
warfarin, 115
Warner, John, 278–282
Warner Babcock Institute (WBI) for Green Chemistry, 278, 280–281
wasps, 23
waste management, 66, 267-269; in balanced biosphere, 276; biomimicry impact on, 21, 173; mushrooms in, 168; in oceans, 290–291
wastewater handling, 104, 162–163
water, 63, 65, 159–160, 283; adhesive properties of, 177; in energy generation, 178–180; pressure waves in, 208; pumping, 185;
Waterman, T. H., 211
water mixing equipment, 60
water striders, 157
Watts, Phil, 100
Weinstock, George, 198
Weinstock, Joel, 116
Westerveld, Jay, 68
Whale, Eric, 190
Whale Heart Satellite Tracking Program, 106
WhalePower, 99–101

whales, 96–106, 208
whirlpools, 39–41, 43, 46, 54, 60, 95, 98, 134
WildThing watercraft, 108, 216, 221
Wilke, Yvonne, 88, 112
wind, 186, 187
window glass, 142–143, 146, 147
wind turbines, 62, 88, 99–101, 180
wine making, 139, 168
wings, 20, 72, 133–135, 144, 145, 154
wireless communications, 109–110
woodpeckers, 129
Woods Hole Oceanographic Institution, 95
worms, 116–117
wound healing, 29, 115, 138, 141, 146, 175, 249
Wright, André-Denis, 241

XAP watercraft, 108, 109

yeasts, 165, 168
Yoon, Sang-Hee, 129

Ziconitide, 209
zoopharmacognosy, 29

ABOUT THE AUTHOR

An award-winning entrepreneur and biomimetic inventor, Jay Harman has taken a hands-on approach to his lifelong fascination with the deep patterns found in nature. In the process, he has founded and grown multi-million-dollar research and manufacturing companies that develop, patent, and license innovative products, ranging from prize-winning watercraft to interlocking building bricks, afterburners for aircraft engines, and non-invasive technology for measuring blood glucose and other electrolytes. He is credited with being among the first pioneering scientists to make biomimicry a cornerstone of modern and future engineering. His latest ventures—PAX Scientific, PAX Water Technologies, PAX Mixer, and PAX Streamline—design more efficient industrial equipment including refrigeration, turbines, fans, mixers, and pumps based on Jay's revolutionary concepts.

Aside from his entrepreneurial exploits, Jay started a boarding school to teach kids about the environment in Australia, became a champion skin diver, sailed his own yacht 27,000 miles on the Indian Ocean, restored the sister ship to Jacques Cousteau's Calypso—and made it his mission to bring the subject of biomimicry to the public. The culmination of Jay's work is the development of "Nature's Streamlining Principle," a guideline for translating nature's extraordinary efficiencies into industrial applications.

Jay's goal—both as an author and an entrepreneur—is to show industry that improving the efficiency of industrial equipment is beneficial for both the bottom line, and the planet.